Nicole Hunziker

# Experimental Studies on Guinea Pig's Eczema

*Their Significance in Human Eczema*

With 37 Figures

Springer-Verlag Berlin · Heidelberg · New York 1969

Privat-docent Dr. NICOLE HUNZIKER

University Clinic of Dermatology (Dir.: Prof. W. Jadassohn)
Laboratory of Prof. hon. E. Bujard (Medical School)
Geneva

ISBN 978-3-642-86674-6     ISBN 978-3-642-86672-2 (eBook)
DOI 10.1007/978-3-642-86672-2

Title-No. 1539

*May I take this opportunity to express my grateful thanks to Prof. W. Jadassohn, for his enthusiastic support and help.*

*Prof. E. Bujard and P. D. Dr. R. Brun are two collaborators to whom I am deeply indebted.*

*I must acknowledge the very expert assistance of Miss P. Gaudin, V. Quinche, N. Baggi and Mrs. D. Vonlanthen.*

*This work was made possible by a subsidy from the Swiss National Fund for Scientific Research.*

# Preface

Dr. Bruno Bloch, professor of the Dermatological clinic of Zurich, began a new era of eczema research by attempting to sensitize guinea pigs to primula. It was during this period that I had the pleasure of being Dr. Bloch's chief resident, and could observe these experiments.

It was only in 1955 that Dr. E. Bujard, professor of Histology of the University of Geneva, Dr. R. Brun, chief of laboratories of my Dermatology Department and I began to work in the field of experimental eczema on the nipples and flanks of guinea pigs. Besides these collaborators, to whom I am very grateful, a few others have participated in our experiments. Among these new co-workers the most important was Mrs. Nicole Hunziker, chief resident of my clinic. From 1960 on, she worked on the eczema experiments with guinea pigs. Our research has been considerably advanced by Dr. Hunziker. Untill now we have published 30 communications on this subject, but the publications about our results have been very short. It now seemed appropriate to publish a monograph and Dr. Hunziker undertook this very toilsome task. I would like to take this occasion to thank her for her many years of collaboration both in the clinic and in the laboratory. I am also very grateful to Professor Bujard who, in spite of the fact that he is retired, continues to work with us. We continually profit from his great experience. Dr. Robert Brun is working with us and always maintains his enthusiasm. I consider myself very lucky to have had such co-workers as Eugene Bujard, Robert Brun and Nicole Hunziker.

Experiments on guinea pig eczema are very important, not only as basic research but also from a practical point of view, for guinea pig eczema has many analogies with human eczema. If Dr. Hunziker's monograph can stimulate new experiments in this domain or facilitate other work in this field, then her efforts may be considered as worthwhile.

Geneva, November 1968                                            Prof. W. Jadassohn

# Table of Contents

# Abbreviations employed in the text

CA  : citraconic anhydride
PA  : propionic anhydride
DNCB : dinitrochlorobenzene
PNDMA : paranitrosodimethylaniline

# Introduction

The first observations on experimental eczema date back to the end of the last century, thus opening the chapter of contact eczema.

Indeed, in 1896, J. JADASSOHN, after applying a mercurial plaster to the arm of a patient, observed for the first time a local reaction. He thus produced experimentally a bullous lesion and he was able to affirm that the eczematous lesions presented by this patient were caused by a hypersensitivity to mercury. This first experiment permitted the separation of contact eczema from the large group of eczemas of indeterminate origin.

The first experimental sensitizations on man were made by NESTLER (1904), CASH (1911) and LOW (1924). BLOCH and STEINER-WOURLISCH (1926 and 1930) were the first to sensitize, using the same method (concentrated extracts of Primula obconica) man and the guinea pig. The latter authors have shown that the histological lesions in man are similar to those of the "spontaneous" eczema caused by Primula and that the histological lesions of the patch test in guinea pigs also show a great similarity to those observed in man.

During the same period, other investigators sensitized guinea pigs with various substances such as nickel (WALTHARD, 1926), arsphenamine (FREI, 1928; SULZBERGER, 1929), and phenylhydrazine (W. JADASSOHN, 1930). At the 8th international Congress of Dermatology and Syphilology in 1931, J. JADASSOHN and BR. BLOCH insisted on defining eczema as an idiosyncratic reaction of the skin.

Since that time the studies concerning experimental eczema of man and animals have greatly increased. Their orientations are different: histological, histochemical and immunological, but they all contribute to the elucidation of the mechanism of contact eczema.

In an attempt to simplify our study, we have permitted ourselves to omit the citing of hundreds of these works and have referred, for publications before 1960, to the bibliographies published by DOERR (1929), W. JADASSOHN (1932), SPIER (1961), MIESCHER (1962), STORCK (1962).

As for the more recent works, we will cite, in the course of this study, all that are pertinent to our research.

# Material and Technique

## A. Choice of Animal

It is neither convenient nor agreeable to experiment with man. Furthermore we cannot foresee what sensitization may lead to, even though the substance one uses may not often be found in everyday life (cross-sensitization?, polyvalent sensitization? etc.). On the other hand, this necessarily limits the number, time and quality of experiments. Nevertheless, the authors who have described experimental sensitizations in man give us interesting points of comparison.

As for animals, sensitization is easiest to induce in monkeys, guinea pigs and pigs. It is more difficult in chickens and dogs and extremely difficult in rats (lit. see STORCK, 1962).

For practical and financial reasons, we have naturally preferred *guinea pigs* to monkeys. Moreover since the classical work of BLOCH and STEINER-WOURLISCH (1930) on Primula eczema in guinea pigs, this animal has been chosen by most investigators working with experimental eczema. Thus we can better compare the results of these various works with our own.

We used male albino guinea pigs: albinos because the macroscopic and microscopic reactions are easier to observe than in pigmented skin; male, in order not to be bothered by the oestrogenic cycle or by gestation. In each experiment we used groups of at least 5 to 6 guinea pigs, often more.

We know that certain hereditary characteristics can play a part in sensitization in guinea pigs (CHASE, 1941) and that there also exists in this animal, as in man, individual differences in the capacity to be sensitized. However, we have deliberately not chosen pure breeds, because we think this will provide a more accurate picture of what occurs in a population. Nevertheless, the groups that we compare in our experiments come from the same stock; all animals are of about the same weight (generally between 250 and 400 g); and they were all fed in the same manner.

Seasonal influence can also play a role, according to MAYER and SULZBERGER (1931) and others (s. p. 5).

## B. Macroscopic Control of Sensitization (Patch Test)

The technical conditions in which patch tests are performed for the specific reaction of sensitization as well as for the primary toxic reaction must be well determined:

A few hours before the application of the substance, we shear the flanks with an electrical shearing machine, then we shave them very delicately with an electric shaver. Some hours after this we apply the substance with a cotton swab on a surface of about 3 cm², spreading it in the most uniform manner possible.

As CHASE (1954) emphasized, all causes of possible irritation to the surface being tested, must be eliminated in order to avoid misinterpretation: scratching of the animals, irritation caused by the method of depilation or by the solvent; olive oil may be irritating in 5% of normal animals (FISHER and COOKE, 1958).

We perform the patch test 8 to 15 days after the end of a sensitization treatment and we interpret the macroscopic reaction generally 12 to 14 hours later, for in sensitized guinea pigs the reaction is best seen at this moment; however, it may appear earlier (see citraconic anhydride p. 31).

The incubation time is shorter if the sensitization is great and the concentration high (CHASE, 1954). However, the period of latency cannot in itself be a criterion for the appreciation of the degree of sensitization (NILZEN, 1952). The duration of the macroskopic reaction persists up to approximately 24 hours it varies with the eczematogenic substance used.

Of course, the intensity of the reaction can be evaluated only if there is a clear difference. The various degrees must be evaluated according to a conventional scale:

LANDSTEINER and JACOBS (1935) adapted a descriptive notation following the exact aspect of the observed reaction. For simplification we adopted a standard scale following the common criteria used by many investigators: LANDSTEINER and CHASE (1940, 1941), NILZEN (1952), CHASE (1954), ROCKWELL (1955), and others.

— no reaction,
± doubtful reaction (some spots of erythema in the area tested),
+ net homogeneous erythema,
+ + intense homogeneous erythema,
+ + + intense homogeneous erythema and edema,
+ + + + intense homogeneous erythema, edema, necrosis.

If only important differences are noted and if the test is always done under the same conditions, the macroscopic interpretation of the patch test is certainly of value.

In order to avoid all risk of misinterpretation, the concentration of the substance used in the patch test must be lower than the minimum toxic concentration of that substance, but sufficient to trigger a reaction in the sensitized guinea pig. This concentration can be lower in the more sensitized animals than in those less sensitized. Guinea pigs have been shown to be sensitive to concentrations 10 to 20 times lower than the minimum toxic concentration (SULZBERGER and WITTEN, 1954). However, man can react to concentrations 100000 times lower than the minimum primary toxic concentration (WEDROFF and DOLGOFF, 1935, and others).

We would like to insist, as have MACHER and SENNLAUB (1963), that the main source of error which leads to misinterpretation lies in the very small difference between the minimum concentration of the solution used in the patch test for the specific reaction of sensitization and the minimum concentration producing a primary toxic effect.

In every laboratory, in fact in all new experiments, these concentrations must be re-evaluated.

## C. Histological Control of Sensitization

Histological examination of the specific reaction of sensitization as well as of the primary toxic reaction can be done either on the flank, at the site of the patch test, or on the nipple. The macroscopic examination of the latter is difficult (the surface being too small), but the histological examination is on the other hand easier to interpret because there are no hair or nearly no follicles and the epidermis is thicker and more closely resembles the human epidermis. For these reasons W. JADASSOHN, BUJARD and BRUN (1955) chose the nipple as the test-organ for their histological study of experimental eczema in guinea pigs. For the macroscopic reaction, we therefore, concerned ourselves only with the flank but we examined the reaction histologically both on the flank and on the nipple.

Biopsies were performed without local anesthesia, and were fixed in Bouin's liquid (mixture of an acqueous solution of picric acid, formaldehyde and acetic acid) or in formol (to show eosinophilic leucocytes), then immersed in paraffin and finaly sliced and stained with hematoxylin and eosin according to routine methods.

In certain cases, we used the alkaline phosphatase reaction according to the method of GOMORI (1941), modified by DANIELLE, and described by LISON (see technic BUJARD, 1957, and N. HUNZIKER, 1964).

1*

Biopsies are usually done approximately 14 hours after the application of the sub-stance, for at this moment the macroscopic reaction is often optimal. As we have already stated, this permits us to compare the histological reactions obtained by various allergens.

In certain cases, however, the biopsies have been successively done between the 2nd and 24th hours and even later after the application of the eczema producing substance, in order to follow the evolution of the primary toxic and that of the eczematous reaction.

# I. Experimentation with Dinitrochlorobenzene (DNCB)

LANDSTEINER and JACOBS (1935) chose 1-2,4 chlordinitrobenzene and other derivatives of benzene for their work on the sensitization of guinea pigs. This eczematogenic substance had already been used by WEDROFF (1933) in man. A very close derivative, O-dinitrophenol, was known to be a frequent cause of allergy in workers handling this substance (FRUMESS, 1934).

As there are 90 chloride or nitrous derivatives of benzene, the question which pre-occupied LANDSTEINER and JACOBS (1935) was that of knowing if they were all sensitizing. Some of these derivatives form in vitro a substitution compound with an organic base, aniline; whereas the other derivatives do not react with aniline. The derivatives which react are the ones capable of sensitizing guinea pigs. This was confirmed in man by SULZBERGER and BAER (1938).

In 1937, LANDSTEINER and CHASE, after having increased their research, deduced from their observations that in order to obtain a sensitization, the chemically active substance must be first combined with an animal protein to form a complete antigen. This hypothesis had already been formulated by WOLF-EISNER in 1907.

Since the works of LANDSTEINER and JACOBS on the sensitization of the guinea pig (1935) and those of WEDROFF and DOLGOFF (1935), SULZBERGER and BAER (1938), HAXTHAUSEN (1939 and 1940), and many others, on the sensitization of man, DNCB is one of the substances most often used in studying the various problems of experi-mental eczema.

## A. Method of Sensitization

### 1. Methods Employed in Our Studies

#### a) *Application of a Primary Toxic Solution*

In most of our experiments, guinea pigs were sensitized by the use of a primary toxic solution (DNCB 1% in acetone).

Before the application of the sensitizing solution on the neck of the guinea pig, the hairs are sheared and then shaved with an electric shaver. The solution is then applied with a cotton-swab every day for 12 days* on a surface of approximately 3 cm in diameter. The degree of sensitization is evaluated 8 to 15 days later.

---

* 11 applications in 12 days.

Most of the guinea pigs are thus sensitized, if not we apply the same solution 3 to 6 times more.

Contrary to what we have observed with other eczematogenic substances, we have never seen guinea pigs resistant to sensitization with DNCB.

Nevertheless Chase (1941) shows that, according to the heredity of the animals, certain guinea pigs are extremely difficult to sensitize. However he adds that: "It is highly doubtful whether an absolute resistance will be encountered in any guinea pig under more intensive treatment, for instance repeated application of an oil or alcohol solution to the skin".

We are under the impression that it is easier to sensitize guinea pigs in the winter. However, we do not wish to draw any conclusions, because, our experiments do not permit us to establish with certainty a periodic factor contributing to sensitization.

Mayer and Sulzberger (1931) using Ursol (paraphenylenediamine) and Salvarsan showed that in winter 75% of the animals are strongly sensitized, whereas in summer only 12% are sensitized.

This difference between winter and summer has been confirmed by Götz and Schulz (1956) for DNCB.

Using a primary non toxic solution of DNCB for sensitization, de Weck and Brun (1956) obtained negative results in summer but their results were positive in winter. These experiments should be repeated.

### b) Application of a Primary Non-Toxic Solution

Guinea pigs can also be sensitized by regular, simultaneous applications of a 0.1% acetonic solution of DNCB (primary non-toxic) on both nipples and both flanks (Frey, 1951).

We have asked ourself the question: *can guinea pigs be sensitized by daily applications of a drop of a 0.1% solution of DNCB on only one nipple?* (N. Hunziker, 1961.)

After 5 daily applications we did not observe a general sensitization (in 5 patch tests, not one was positive).

Table 1. *Sensitization with applications of a primary non-toxic solution of DNCB on the nipple of guinea pigs*

| Number of daily applications of a 0.1% solution of DNCB (in acetone) on the nipple | Number of guinea pigs (male albinos, weight approx. 250 g) | Epicutaneous test of DNCB 0.1% (in acetone) on the guinea pigs side (reading taken after 14 hours) | | |
|---|---|---|---|---|
| | | + | ± | — |
| 5 | 5 | 0 | 0 | 5 |
| 8 | 10 | 5 | 0 | 5 |
| 20 | 23 | 21 | 2 | 0 |
| 40 | 13 | 13 | 0 | 0 |

+ Distinct homogeneous erythema.
± Doubtful reaction, a few erythema-like spots disseminated on the tested region.
— No reaction.

After 8 daily applications we observed a clear sensitization in some guinea pigs, demonstrated by the patch test on the flank (out of 10 tests, 5 were positive). After 20 to 40 applications on a single nipple, the patch tests on the flanks are positive in almost all of the guinea pigs. Therefore, a general sensitization occurs even though the multiple applications are performed on a very small surface and with a primary non-toxic solution of DNCB (Table 1).

## 2. Other Methods

### a) Sensitization by Epidermal Applications

As in a certain number of our experiments, some investigators used primary toxic solutions of DNCB*; others employed primary non-toxic solutions of DNCB**, as we did in some of our experiments. There are many technical differences (concentration, number of applications) in addition to numerous variations in the results obtained by experimentation on guinea pigs in different laboratories as well as in the same laboratory.

The success of sensitization does not depend only upon the quantity of the allergen applied, but also upon the concentration of the substance per surface unit (SCHNITZER, 1941; MIESCHER, 1941; FREY and WENK, 1956).

### b) Sensitization by Injections

*By intradermal administration.* LANDSTEINER and JACOBS (1935) practiced repeated injections of 0.0025 mg of DNCB in an alcoholic solution mixed with physiological serum; this procedure regularly induced a sensitization. The guinea pigs thus treated react to the patch test in 0.1% concentration of DNCB in olive oil (SKOG, 1955) and even in 0.04% (CHASE, 1941).

Other authors confirmed this means of sensitization (GINSBERG and coll., 1939; ROSTENBERG, 1947; ROCKWELL, 1955; MAIBACH and MAGUIRE, 1963; REBELLO and SUSKIND, 1963).

*By sub-cutaneous or intramuscular administration.* LANDSTEINER and JACOBS (1935) and SEEBERG (1951) obtained a weak and irregular sensitization. Whereas MIESCHER (1941), then SCHNITZER (1942), regularly obtained a sensitization by intramuscular administration of DNCB.

*By intraperitoneal administration.* Positive results are rare (LANDSTEINER and CHASE, 1940).

*By intraganglional administration.* SEEBERG (1951) sensitized a certain number of animals but the concentration of the solution of DNCB used in this test is high (1%), see p. 7 primary toxic reaction.

GRIMMER (1961), obtained a sensitization in all guinea pigs, by injections in the ganglions of the popliteal fossa of an alcoholic solution of DNCB (900 $\gamma$) or of a suspension of pure cristals of DNCB (400 to 500 $\gamma$). The histological examination of the tests on his guinea pigs (p. 70) is particularly interesting.

---

* LANDSTEINER, JACOBS (1936), SCHREIBER and MÜLLER (1938), ANKE (1939), GINSBERG and coll. (1939), HALTER (1941), HÜLLSTRUNG and HACK (1941), SCHNITZER (1942), HAXTHAUSEN (1943), ROSTENBERG (1947), KALKOFF (1948), HÖSLI (1948), GENTELE and HOLMGREEN (1951), SEEBERG (1951), NILZEN (1952), SEEBOHM and coll. (1954), ROCKWELL (1955), JADASSOHN, BUJARD, BRUN (1955), GERSHBEIN and coll. (1956), FREY, WENK (1956), DE WECK, BRUN (1956), BAER and coll. (1957), RAJKA, HARD (1960), DE GRACIANSKY and coll. (1960), GRIMMER, SPIER (1961), RAAB (1962), MACHER (1962).

** SCHNITZER (1942), HÖSLI (1948), FREY (1951), SEEBERG (1951), SCHEPANK (1955), ROCKWELL (1955), DE WECK, BRUN (1956), DE GRACIANSKY and coll. (1960).

MACHER and SENNLAUB (1963) also injected DNCB into the lymph nodes (0.1% to 5% or 65 $\gamma$ to 3550 $\gamma$ per guinea pig). They obtained 50 sensitized guinea pigs out of 118 (solution used for the test: 0.1 and 0.25% DNCB in alcohol).

*By splenic administration.* GRIMMER (1961) obtained a sensitization by injecting in the spleen 900 $\gamma$ of DNCB in alcohol.

*By intravenous or intracardiac administration.* This method is of no interest (LANDSTEINER and JACOBS, 1935; LANDSTEINER and CHASE, 1940).

### c) Sensitization by Injections of DNCB Combined with Epicutaneous Applications of a DNCB Solution

Intradermal injections, combined with the application of a DNCB solution on the skin of guinea pigs, induce an appreciable sensitization (LANDSTEINER and CHASE, 1937). This technique has been produced by many authors, either with DNCB, or with other allergens. We have also employed it for citraconic anhydride (p. 32) and propionic anhydride (p. 35).

### d) Complex Methods of Sensitization

These methods have been especially developed by LANDSTEINER and JACOBS (1936), LANDSTEINER and CHASE (1940, 1941) and CHASE (1954): There are 3 preparations for the injection:

1. Antigen (dinitrochlorobenzene or better dinitrofluorobenzene) + protein = conjugated.
2. Dinitrochlorobenzene + Freund's adjuvant.
3. Complete antigen (conjugated) + adjuvant.

FISHER and COOKE (1958) employed the combined method of sensitization according to CHASE (1954) and CHASE and BATTISTO (1955).

KLASCHKA (1966) successfully caused sensitization by intraperitoneal administration, using the complex method described by LANDSTEINER and CHASE (1937, 1940).

# B. Primary Toxic Reaction in Unsensitized Guinea Pigs

Determination of the minimum primary toxic concentration of DNCB in guinea pigs is absolutely necessary before undertaking the sensitization and the interpretation of the tests. In fact, there are important differences of opinion in the literature concerning the primary toxic reaction of DNCB.

Different observations are perhaps due to the fact that the guinea pigs were tested in different laboratories (race, color, weight). CHASE (1954) himself has shown differences in concentration for the primary irritation of his guinea pigs. DE WECK and BRUN (1956) have also observed individual differences in our laboratories.

In our experiments on the primary toxic reaction, we used a 1% solution of DNCB in acetone, which gives a well defined erythematous reaction after 14 hours (patch test for a primary toxic reaction).

### 1. Evolution of Lesions.
### Histological Examination of Primary Toxic Lesions

#### a) Flank

After 2, $3^1/_2$, and 5 hours, we did not observe epidermal or dermal lesions, (except a slight infiltration).

After 14 and 24 hours we observe sub-epidermal bullae, a more or less extended necrosis of the epidermis and a strong dermal infiltration composed of leucocytes and lymphocytes.

### b) Nipple

After 2, $3^1/_2$ and 5 hours, there is very little change, except a slight infiltration and some lymphocytes in the epidermis, particularly visible in the reaction with alkaline phosphatase in the form of black spots (Fig. 1).

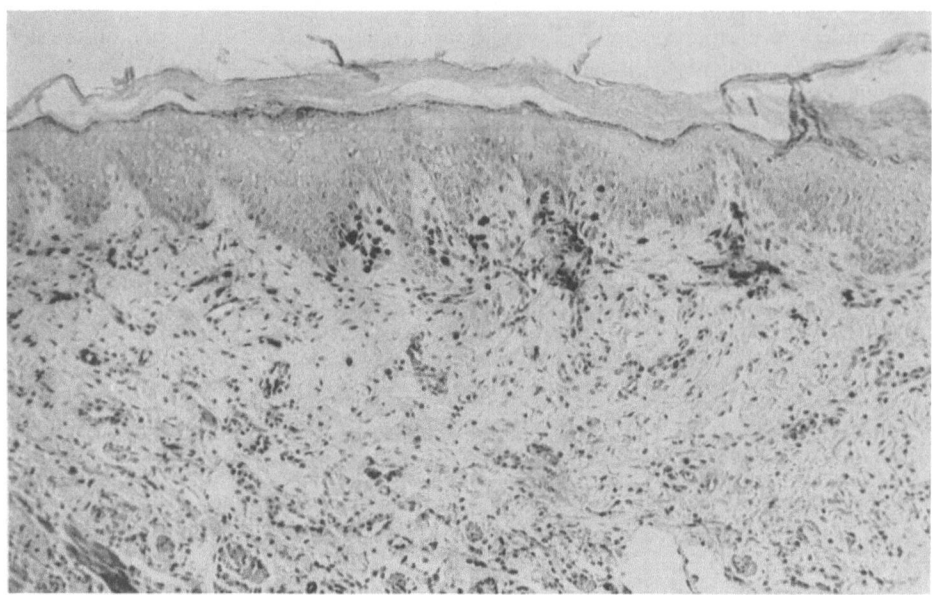

Fig. 1. Non-sensitized guinea-pig. Nipple $3^1/_2$ hours after a single application of 1% DNCB in acetone (alkaline phosphatase): lymphocytes in the superficial part of the dermis

After 6 hours, W. JADASSOHN, BUJARD and BRUN (1955) observed the formation of subepidermal bullae, (the cells of the top of the bullae still being normal), (Fig. 2). After 12 hours these cells are very much altered, (Fig. 3), and some hours afterwards, there are no more intact epidermal cells. After 24 hours the epidermis is completely necrosed.

### 2. The Primary Toxic Effect and Acanthosis

W. JADASSOHN (1944) has already proved the possible protective effect of acanthosis against the toxic reaction of chrysarobine. A certain protective effect of acanthosis was demonstrated by DE WECK and BRUN (1957) against the primary toxic action of DNCB.

### a) Flank

In order to obtain an acanthosis of the epidermis of the flank, a non-greasy cream* is applied 8 times within 10 days on the flank which has been previously shaved; this causes significant acanthosis of the epidermis.

---

* Cream 214: cetyl alcohol 4 parts, propyl alcohol 2 parts, isopropyl myristate 2 parts (BUJARD, BRUN and W. JADASSOHN, 1957).

In unsensitized guinea pigs, a primary toxic solution of 1% and 2% DNCB in acetone is applied on the acanthosed flank as well as on the normal flank.

In applying a 1% solution of DNCB on the flank without acanthosis, the primary toxic lesions already described are observed after 24 hours (sub-epidermal bullae, necrosis of the epidermis, infiltration), whereas on the flank with acanthosis there are practically no lesions (Fig. 4a and b).

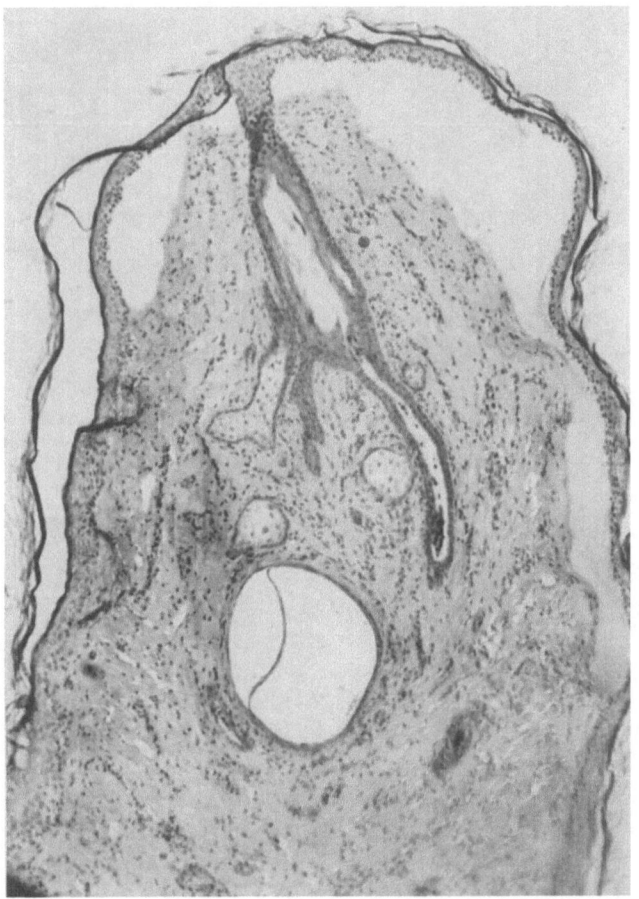

Fig. 2. Non-sensitized guinea-pig. Nipple 6 hours after a single application of 1% DNCB in acetone (hemalun-eosin). Detachment of the epidermis

However if a 2% solution of DNCB is applied on the flank with acanthosis, a primary toxic reaction is also observed (sub-epidermal bullae, necrosis), but to a lesser extent than on the flank without acanthosis.

### b) Nipple

W. JADASSOHN, UEHLINGER and FIERZ (1941) demonstrated by their experiments on the "nipple test", that the application of oestrogen hormones on the nipples of male guinea pigs provokes an intense acanthosis of the epidermis.

For this reason we applied daily for a period of 11 days, a solution of Hormoestrol (2$\gamma$), a synthetic oestrogen, which permitted us to obtain a very strong acanthosis (epidermis 4 to 5 times thicker than that of untreated nipples).

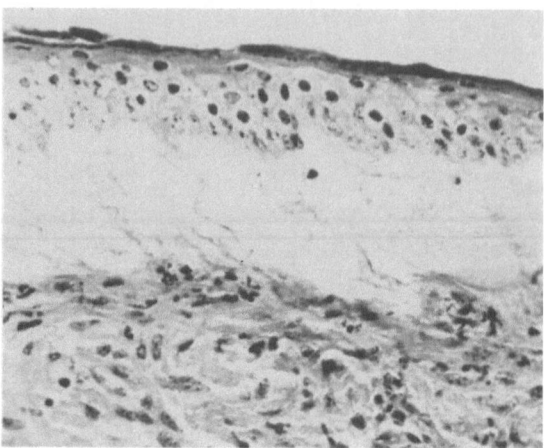

Fig. 3. Non-sensitized guinea-pig. Nipple 12 hours after a single application of 1% DNCB in acetone (hemalun-eosin). Detachment of the epidermis

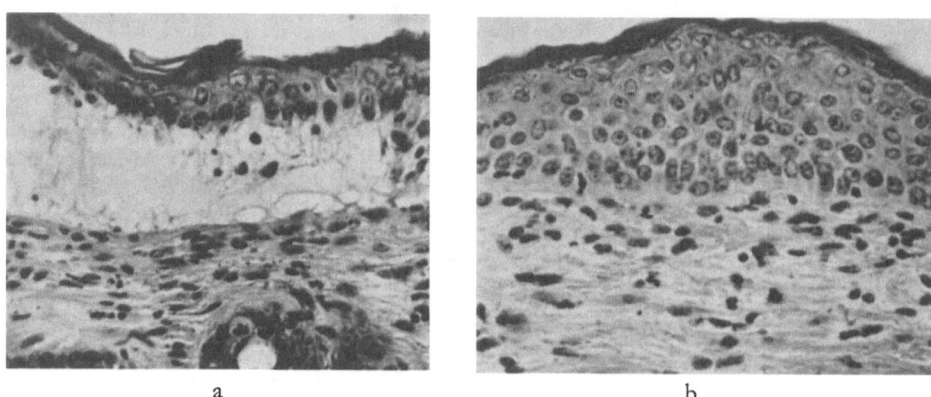

a                                                                 b

Fig. 4a and b. Application of 1% DNCB on the flank of non sensitized guinea pig. a No acanthotic flank: strong alterations. b Acanthotic flank (acanthosis induced by erucic acid): no reaction

The application of Hormoestrol in this dosage on one nipple also provokes a certain degree of acanthosis in the other nipple (hematogenic effect), which makes it impossible to compare the reactions of the 2 nipples of the same guinea pig.

If a 1% solution of DNCB is applied on the nipple with acanthosis, there are no lesions (excision 24 hours after application). However if the same solution is applied on a nipple without acanthosis, the usual primary toxic lesions are observed (necrosis of the epidermis).

*In conclusion, acanthosis protects against the primary toxic action of DNCB, but only to a certain extent, for if the concentration of the solution is increased to 2%, the protection is greatly reduced.*

### 3. The Primary Toxic Reaction and Mitosis

According to most authors, and our studies, a 0.1% primary non-toxic solution of DNCB in acetone causes no macroscopically visible reaction and no histological alterations on the flank and the nipple of guinea pigs.

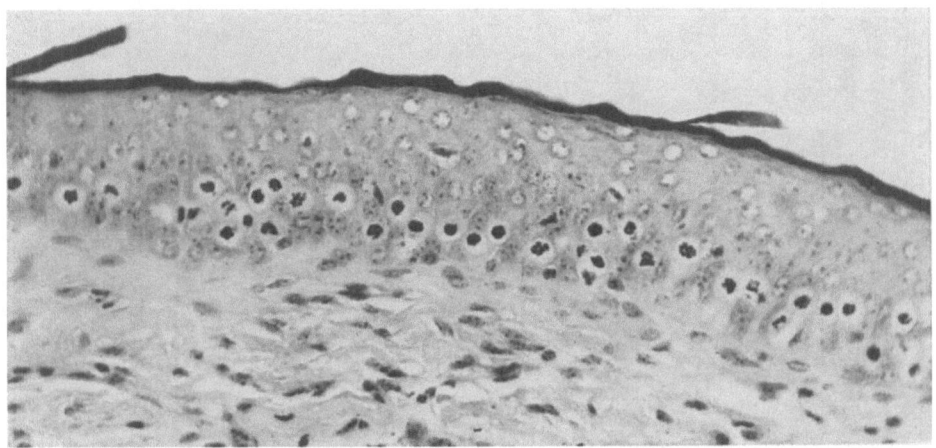

Fig. 5. Non-sensitized guinea-pig. Nipple 33 hours after a single application of 0.1% DNCB in acetone (colchicine injected 9 hours before the excision): Mitosis

BUJARD, W. JADASSOHN, BRUN and PAILLARD (1953) and W. JADASSOHN, BUJARD and BRUN (1955) showed that if colchicine ($50\gamma/100$ gr) is injected into guinea pigs (Dustin reaction) 12, 24 and 48 hours after the test with DNCB, and if the nipple is excised 9 hours later (that is to say in the 21st, 33rd and 57th hour) a temporary increase of mitosis is noted. The increase is not yet visible in the 21st hour, is optimal in the 33rd hour (Fig. 5), and disappears in the 57th hour.

Mitotic activity can also be observed with certain eczematogenic derivatives of benzene (BUJARD, W. JADASSOHN, BRUN and PAILLARD, 1953) (Table 2).

These experiments show that *a substance dissolved in a certain solvent (at a certain concentration) cannot be said to be ineffective, because only a histological examination could reveal certain lesions such as the increase of mitosis.*

The primary toxic reaction of DNCB has been histologically examined by various authors. Among others:

NILZEN (1952) admits that the primary irritation caused by DNCB in unsensitized guinea pigs does not differ from the specific reaction of sensitization; in any case, there is no clear difference.

Table 2. *Mitotic activity of certain eczematogenic derivatives of benzene*

| Substances examined | F, NO$_2$, NO$_2$ | Cl, NO$_2$, NO$_2$ | Br, NO$_2$, NO$_2$ | Cl, NO$_2$ | Cl, NO$_2$, Cl | Cl, Cl | O$_2$N, Cl, NO$_2$ | O$_2$N, OH, NO$_2$ |
|---|---|---|---|---|---|---|---|---|
| Effect on the epidermis of a guinea pig's nipple Sol. 0,1 % | M.D.[a]: 3.8 | M.D.: 3.4 | M.D.: 4.25 | M.D.: 1.6 | M.D.: 1.25 | M.D.: 1.5 | M.D.: 1.2 | M.D.: 1 |
| id. sol. 1 % | necrosis of the epidermis | necrosis of the epidermis | necrosis of the epidermis | M.D.: 0.25 | M.D.: 1.0 | M.D.: 0.5 | M.D.: 3.1 | M.D.: 0.8 |
| Sensitizing effect on the guinea pig (Landsteiner and Jacobs) | + | + | + | − | − | − | + | — |
| Sensitizing effect on man (Sulzberger and Baer) | + | + | | − | | | | |

For every dilution of each substance, 5 guinea pigs were employed.

[a] The mitotic density (M.D.) was estimated by means of an arbitrary scale calibrated from 0 to 6. The numbers in this table have no absolute value and are the arithmetic average of evaluations performed on the 5 guinea pigs.

When the epidermis of the nipple showed necrosis, it was not possible to make an evaluation.

ZELIGMAN (1954) showed that 0.1% DNCB in vaseline in unsensitized guinea pigs produced (after several applications until the appearance of an erythema) an acanthosis* with hypergranulation and hyperkeratosis, an intercellular edema and the formation of vesicles. One percent DNCB in acetone, after one application, provokes spongiosis, intercellular edema, exocytosis, and a moderate acanthosis. Five percent DNCB in acetone provokes a necrosis of the epidermis.

FISHER and COOKE (1958) observed the development of a primary toxic reaction by applying, on the skin of normal guinea pigs, DNCB in olive oil in concentrations of 2%, 3% and 5%.**

# C. Specific Reaction of Sensitization

## ("Eczematous Reaction", "Delayed Reaction")

From the works of various authors*** using DNCB, we note great differences in the concentration of the solution of DNCB employed.

According to results obtained by various authors and ourselves, the concentrations sufficient to give an interpretable patch test, but hardly ever a primary toxic reaction, are for DNCB solutions:

|  |  |
|---|---|
| in acetone | 0.1%, |
| in alcohol | 0.1%, |
| in olive oil | 0.5%. |

As for the primary toxic reaction, not only the concentration of the allergen, but also the solvent, must be taken into consideration for the test. Solutions of DNCB diluted in oil are less potent than alcoholic or acetonic solutions.

NILZEN (1952) obtained, with 0.1% DNCB in acetone, 100 positive tests out of 120 sensitized guinea pigs; whereas with a solution of 0.1% DNCB in olive oil, there were no positive tests in the same guinea pigs. However, when the concentration of DNCB in oil was increased to 1%, all of the tests were positive (120 out of 120).

## 1. Patch Test

*In all of our experiments, we used, for the patch test, a solution of 0.1% DNCB in acetone.* In the macroscopic and histological examinations of unsensitized guinea pigs this solution does not give any positive reaction but merely a temporary in-

---

* Nevertheless GAUDIN (1948) in our clinic showed that repeated applications of vaseline alone can provoke an acanthosis. This was confirmed by SCHAAF and GROSS (1953), SCHULZ (1962) and many others.

** According to these experiments, a single application of olive oil on the flank occasionaly causes a vacuolisation of the superficial epidermal cells, as is also seen sometimes in regions shaved with an electrical shaver, even without the application of olive oil.

*** LANDSTEINER, JACOBS (1935), SCHREIBER, MÜLLER (1938), HÜLLSTRUNG, HACK (1941), LANDSTEINER, CHASE (1941), CHASE (1941), MIESCHER (1941), SCHNITZER (1942), HAXTHAUSEN (1943), ROSTENBERG (1947), ROSTENBERG, HAEBERLIN (1950), FREY (1951), FREY, STUDER (1951), SEEBERG (1951), HAXTHAUSEN (1951), GENTELE, HOLMGREEN (1951), NILZEN (1952), NILZEN, HUSSEY (1954), SEEBOHM, coll. (1954), CHASE (1954), W. JADASSOHN, coll. (1955), SKOG (1955), ROCKWELL (1955), GERSHBEIN, coll. (1956), DE WECK, BRUN (1956), BAER, coll. (1956), BAER, coll. (1957), FISHER, COOKE (1958), MIESCHER (1962), KLASCHKA (1964).

crease in mitotic activity (see page 11). In sensitized guinea pigs, on the contrary, one observes an erythema, localized at the point of application of the DNCB solution, which appears after 6 hours, increases progressively but diminishes after 24 hours.

### 2. Histological Examination of the Eczematous Reaction

#### a) Flank

Twelve, 14 and 24 hours after the application of 0.1% DNCB in acetone, we often observe lesions of an eczematous aspect: outlines of spongiosis, clear spongiosis, (Fig. 6), occasionally vesicles, cells of the infiltrate invading the epidermis, sometimes thickening of the epidermis (acanthosis), and a dermal infiltrate composed largely of lymphocytes, with few or no eosinophilic leucocytes

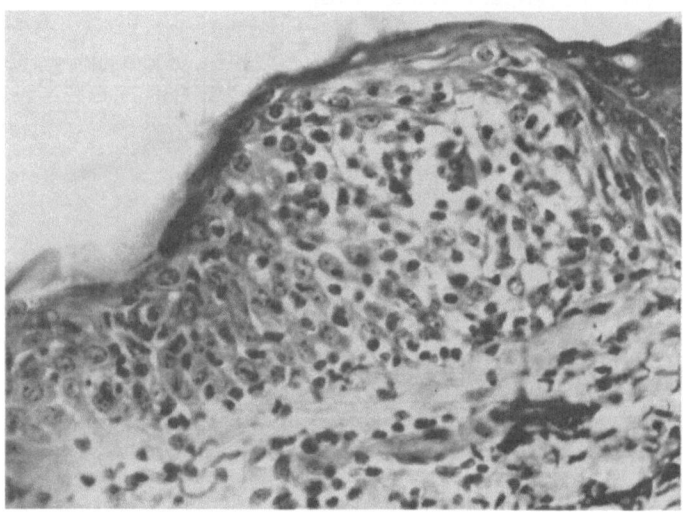

Fig. 6. Sensitized guinea-pig. Reaction seen on the flank 24 hours after a single application of 0.1% DNCB in acetone: Spongiosis

It should be noted, however, that the histological lesions, especially the spongiosis, are not always seen. This is probably due to the thinness of the epidermis of the guinea pig's flank (2 to 3 cellular layers).

In certain cases it is very difficult, according to the appearance of the lesions, to ascertain whether it is an eczematous reaction (0.1% DNCB in acetone) in sensitized guinea pigs or a primary toxic reaction (1% DNCB in acetone) in unsensitized guinea pigs (NILZEN, 1952, p. 11 and ZELIGMAN, 1954, p. 12). One should keep in mind that a 0.1% DNCB solution in acetone does not provoke histological lesions in unsensitized guinea pigs.

#### b) Acanthosis of Flank

In guinea pigs sensitized with DNCB, the reactions obtained on the flank with acanthosis* are more typical and easier to interpret than those obtained on the flank

* For the procedure for provoking an acanthosis, see p. 8.

without acanthosis (BAER and coll., 1956; DE WECK and BRUN, 1957; REBELLO and SUSKIND, 1963).

We have repeatedly observed on the flank with acanthosis, in addition to spongiosis (Fig. 7), the presence of "altérations cavitaires", already described in 1890 by LELOIR as typical of eczema in man.

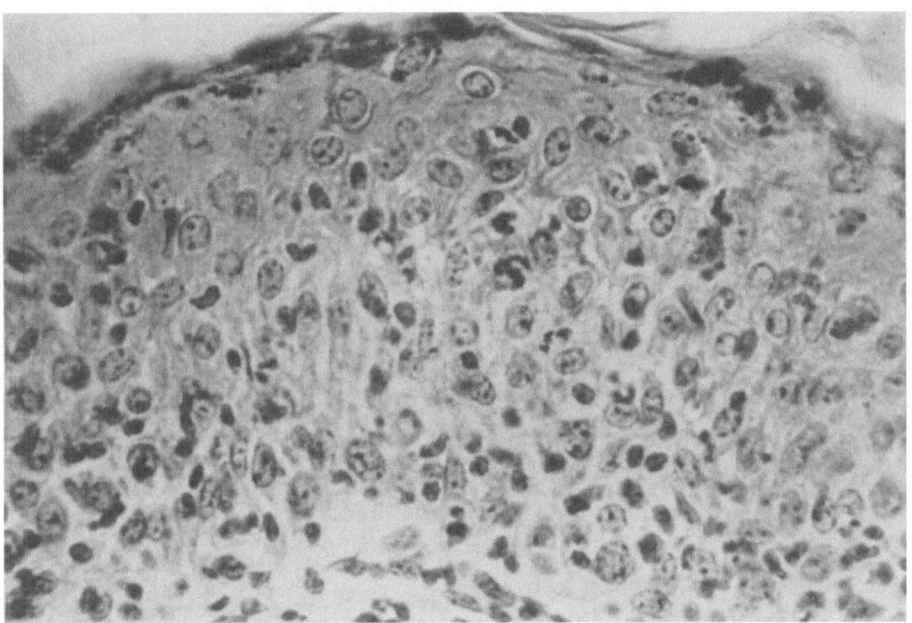

Fig. 7. Sensitized guinea pig. Reaction on the acanthotic flank 14 hours after a single application of 0.1% DNCB in acetone: spongiosis

As UNNA (1894) did in man, we observed on the acanthosed more often than on the unacanthosed flank of guinea pigs, small intraepidermal abcesses filled with polynuclear cells.

### c) Nipple: Evolution of Lesions

*Two and $3^1/_2$ hours after* the test, we already observe outlines of spongiosis or some small true spongiosis*.

With the alkaline phosphatase reaction, this spongiosis does not yet seem to be invaded by the small lymphocytes (round cells) which appear on slides as "black spots". Lymphocytes are disseminated in the dermis and the capillaries. Sometimes they are aligned underneath the epidermis, at the very place where there is already a spongiosis or the beginning of spongiosis (Fig. 8).

---

* This time of $2^1/_2$ to 3 hours is remarkably short but W. EPSTEIN and KLIGMAN (1959) observed in man eczema like lesions already 1 to 2 hours after the test. BAER and coll. (1956) observed the first signs after 5 to 6 hours on the acanthosed flank of guinea pigs.

FISHER and COOKE (1958) have previously observed signs of eczema very early in sensitized guinea pigs by a method according to CHASE (2 hours following the test).

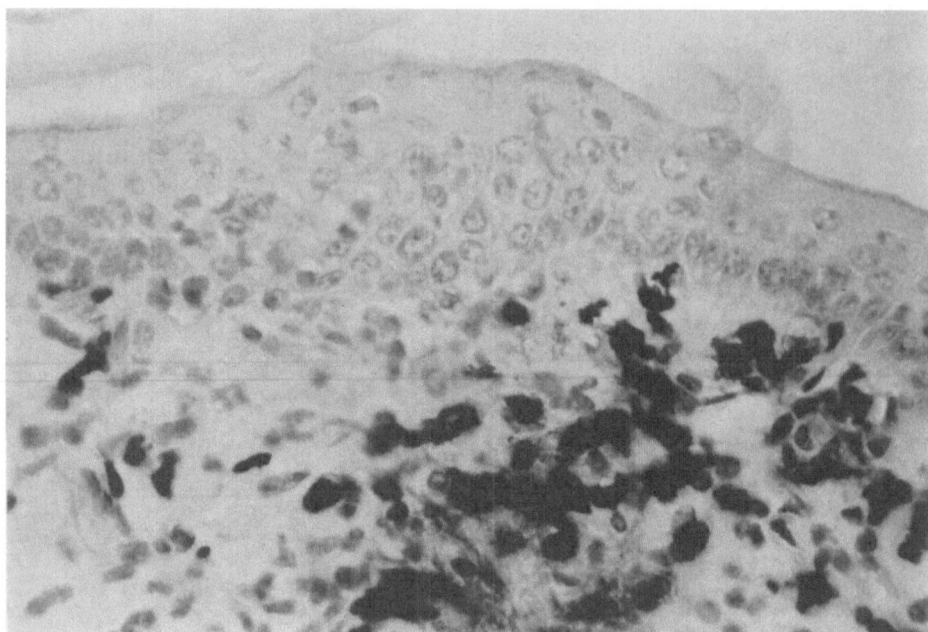

Fig. 8. Sensitized guinea pig. Nipple $3^1/_2$ hours after a single application of 0.1% DNCB in acetone: beginning of spongiosis, dermal infiltration of cells giving a positive reaction with alkaline phosphatase

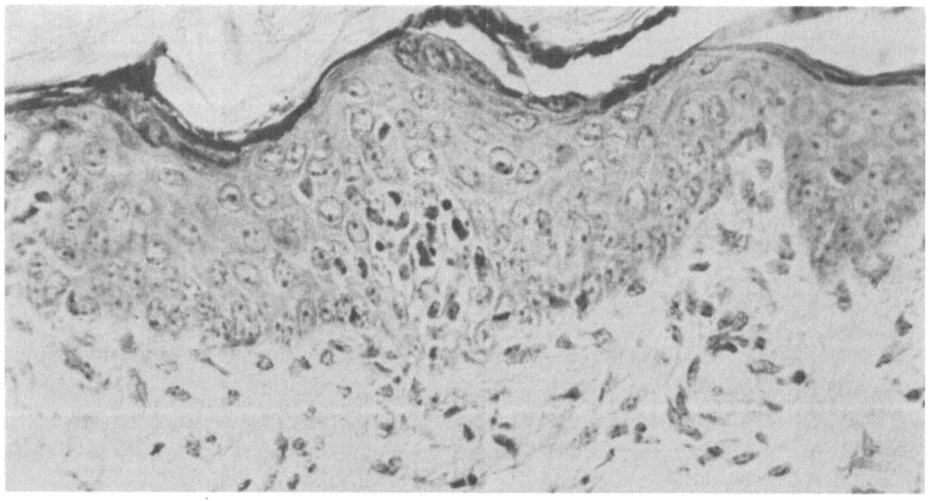

Fig. 9. Sensitized guinea-pig. Nipple 6 hours after a single application of 0.1% DNCB in acetone: spongiosis; little dermal infiltration

*After 5 to 6 hours* spongiosis is clearly visible (Fig. 9). The infiltrate is sometimes minimal and sometimes considerable. The alkaline phosphatase reaction shows on the same slide both spongiosis invaded by lymphocytes ("inhabited" spongiosis) and spongiosis not containing cells of the infiltrate ("non-inhabited" spongiosis).

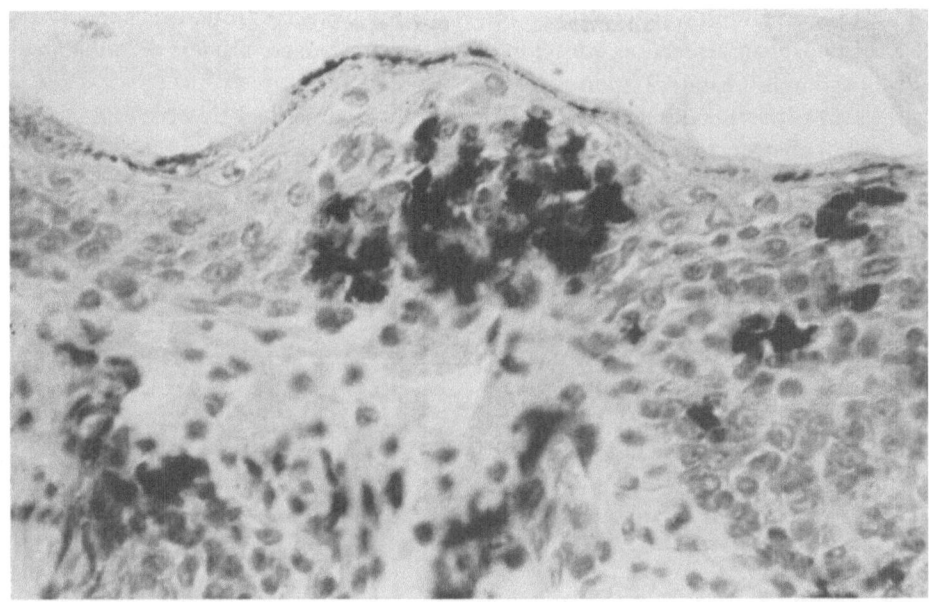

Fig. 10. Sensitized guinea-pig. Nipple 9 hours after a single application of 0.1% DNCB in acetone: spongiosis and cells giving a positive reaction with alkaline phosphatase

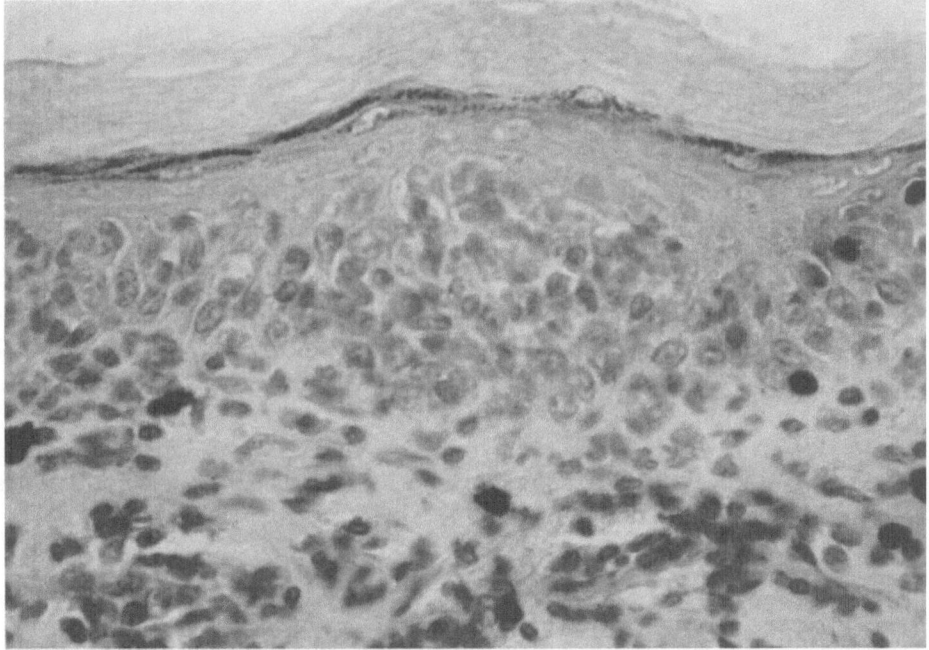

Fig. 11. Sensitized guinea-pig. Nipple 9 hours after a single application of 0.1% DNCB in acetone: spongiosis, but no cells giving a positive reaction with alkaline phosphatase

*After 9 hours* there are numerous clusters of spongiosis in the epidermis and a very clear dermal infiltrate composed largely of lymphocytes (round cells). With the alkaline phosphatase reaction, we again observe that there are clusters of "inhabited" as well as "non-inhabited" spongiosis (Figs. 10 and 11).

*After 14 hours* besides an extensive spongiosis, as described by MIESCHER (1962) (Fig. 12), and sometimes a spongiosis in clusters as in human eczema, we note

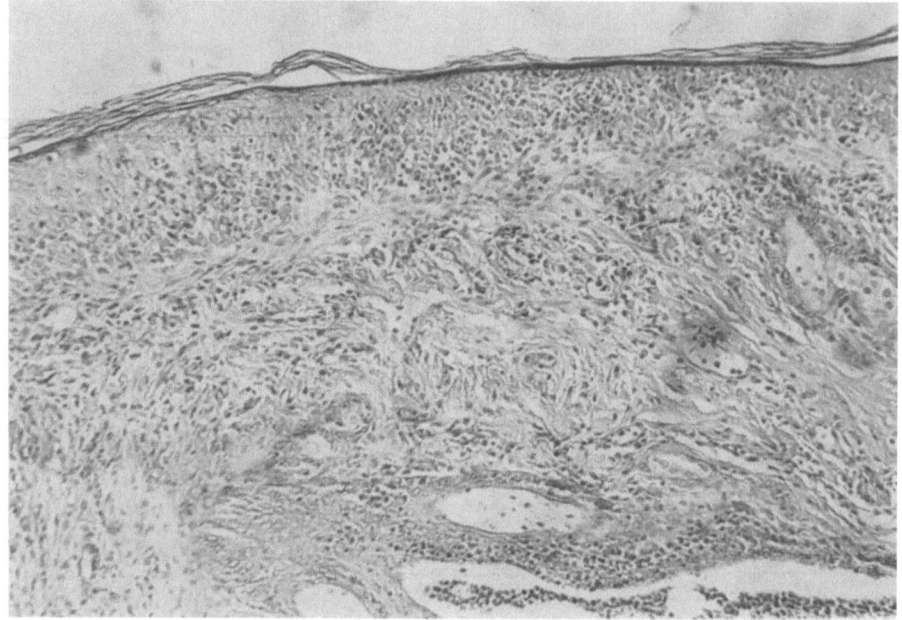

Fig. 12. Sensitized guinea pig. Nipple 14 hours after a single application of 0.1% DNCB in acetone: extensive spongiosis

subcorneal, vesicles extending more or less throughout the superficial half of the epidermis, (Fig. 13), and rarely the small subcorneal vesicles described by UNNA (1894) and by CIVATTE (1925, 1950, 1954). The dermal infiltrate is of considerable importance.

With the alkaline phosphatase reaction, the spongiosis is shown to be totally or partially "habited". The lymphocytes are localized especially in the superior part of the epidermis, particularly in the vesicles. At the dermal level, the cells of the infiltrate vary in number.

*After 21 hours* the increase in spongiosis gives the appearance of intra-epidermal vesicles (W. JADASSOHN, BUJARD and BRUN, 1955). These histological lesions correspond to those observed in man.

*Note:*

Thus, already $2^1/_2$ to 3 hours after the test, the alkaline phosphatase reaction reveals lymphocytes disseminated in the dermis and at times aligned underneath the epidermis. These lymphocytes look like "black spots" on the slides. However, it is difficult to say if it is only the lymphocytes which appear thus and whether all the lymphocytes always give a positive reaction.

GENTELE and HOLMGREEN (1951) when testing sensitization observed at the point of injection of the DNCB solution, a positive reaction with alkaline phosphatase, apparently limited to the cells of the infiltrate.

In human eczema, positive reactions with alkaline phosphatase have similarly been observed in the cells of the infiltrate (FISHER and GLICK, 1947; PIRILÄ and ERÄNKÖ, 1950; KOPF, 1957) and in the capillaries of the papillary layer (NISHIVAMA, 1963).

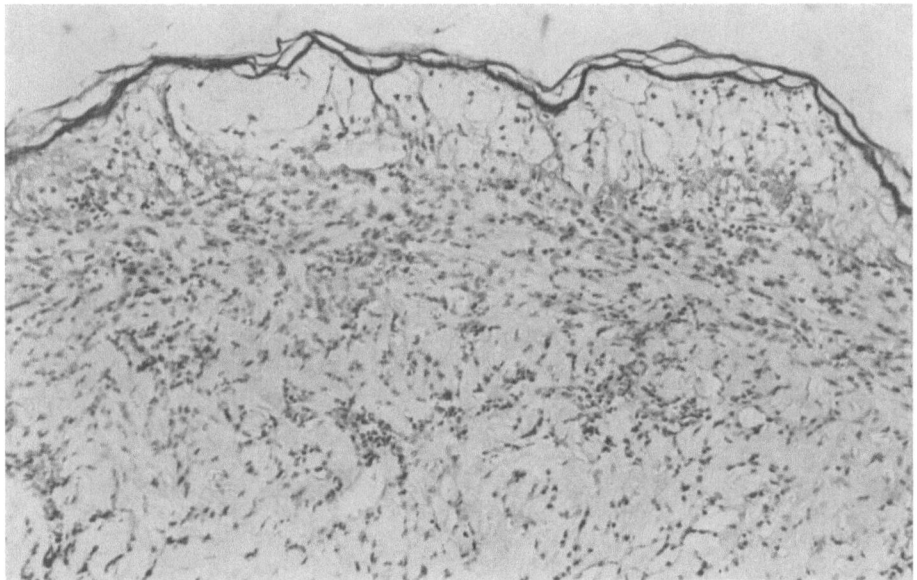

Fig. 13. Sensitized guinea pig. Nipple 14 hours after a single application of 0.1% DNCB in acetone: intraepidermic vesicles, blisters

### d) Acanthosis of the Nipple

On the acanthotic nipples of sensitized guinea pigs the reactions are typical, as previously described: (spongiosis, vesicles, see p. 18). The reactions do not differ qualitatively from those observed in non-acanthotic nipples. They are perhaps less intense and do not lead to significant vesicle formation.

It seemed also (DE WECK and BRUN, 1957) that the course of the reaction was slightly modified by the acanthosis. After 24 hours the lesions on the acanthotic nipples were clearly less significant. "Nonetheless, the histological differences observed between the reactions on the flanks and nipples with or without acanthosis do not permit us to affirm that acanthosis plays a protective role against the triggering of an eczematous reaction in animals sensitized with DNCB...". But we should keep in mind that under certain conditions, acanthosis can have a protective effect relative to the primary toxic reaction (see p. 8).

### e) The Specific Reaction of Sensitization and Mitosis

We have seen that one can demonstrate an increase in mitotic activity on the nipple after application of 0.1% DNCB to unsensitized guinea pigs (see p. 11). This

2*

increase in mitotic activity appears only 33 hours after application and is not lasting, as it disappears 24 hours later.

Using the same technique in sensitized guinea pigs (see p. 11), we also observe an increase of mitotic activity in the epidermis of the nipple 33 hours after the test, but this mitotic increase is still apparent 129 hours after the test (W. JADASSOHN, BUJARD and BRUN, 1955) (Fig. 14).

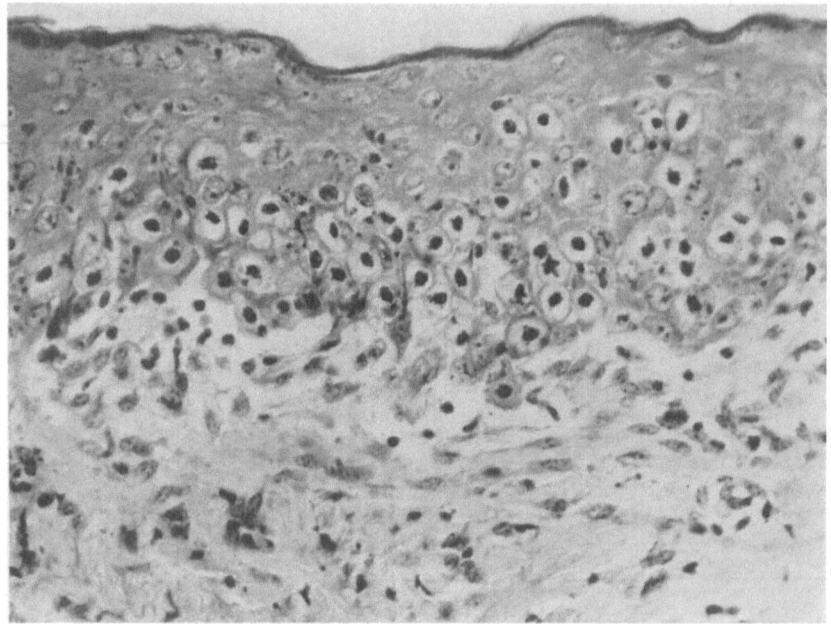

Fig. 14. Sensitized guinea pig. Nipple 33 hours after a single application of 0.1% DNCB in acetone (colchicine injected 9 hours before excision): widespread mitosis

## D. The Histological Lesions of the Epidermis and Dermis

### 1. Spongiosis

We will not repeat the description of spongiosis since this has been described in the evolution of the lesions of the flank and nipple (see p. 14 and 15).

### 2. Intra-Epidermal Vesicles

The intra-epidermal vesicles that we have observed in guinea pigs (Fig. 13) resemble those seen in man. They spread throughout the superior half of the epidermis, but they can completely invade the epidermis. LEVER (1961) writes in his book on histopathology: "In acute dermatitis intra-epidermally located vesicles or bullae dominate the histologic picture. Consider-able inter-cellular edema and spongiosis and intra-cellular edema (altération cavitaire) may be present in the epidermis surrounding the vesicles. If the number of vesicles is great and the intra-cellular edema is pronounced, the vesicles, due to reticular degeneration of the epidermis, will be separated from one another only by thin septa by the resisting walls of edematous epidermal cells and thus will form a multilocular bulla".

This description of human eczema by LEVER (1961) corresponds to the description of the bullae and vesicles observed in the eczema of the guinea pig's nipple.

MIESCHER (1962) also describes guinea pigs strongly sensitized to DNCB, as having the superficial part of the epidermis showing edematous epidermal cells, being discolored, and having no or picnotic nuclei ("pseudospongiotische Aufblähung").

### 3. Subcorneal Vesicles

In 4 adult guinea pigs strongly sensitized by our usual procedure, we noted some small subcorneal vesicles like those described by UNNA (1894) and CIVATTE (1925).

We then re-examined the histological preparations of the 70 guinea pigs sensitized to DNCB and tested in the usual manner; we found no small subcorneal vesicles either on the flank or on the nipple. Hence, these small vesicles are rare in the eczema of the adult guinea pig. On the other hand, we have seen them rather frequently in new-born sensitized guinea pigs. (Fig. 15).

### 4. Dermal Infiltrate

The infiltrate is composed of lymphocytes and especially lymphocytes which invade certain parts of the epidermis and are thus found in spongiosis. In the vesicles, there are leucocytes and lymphocytes, but we detected very few, if any, eosinophilic leucocytes.

FISHER and COOKE (1958), using special staining techniques to show mastocytes, basophiles and eosinophiles, found none of these cells in excess in the biopsies of the patch tests in guinea pigs sensitized to DNCB. However, NIEBAUER (1962), using the toluidin blue stain, noted an unusually large number of mastocytes as well as degenerating mastocytes (Ghost cells).

## E. Sensitization of New Born Guinea Pigs (Small Subcorneal Vesicles)

In 1929, J. FREUND showed that experimentally infected young guinea pigs, had little or no reaction to an intradermal injection of an aqueous extract of tuberculus bacili, despite the presence of large tuberculous lesions in the lymphatic ganglions and in the spleen. On the contrary, the controls, adult guinea pigs infected in the same manner, reacted to the same extract by redness, edema, and necrosis at the site of injection.

A discrepancy between cutaneous and systemic reactions to tuberculin has already been observed in young guinea pigs (VALTIS and SAENZ, 1928).

Numerous accounts in the literature show that the immunological behaviour of young children and animals is different from that of adults (see literature: FREUND, 1929, and immunotolerance, p. 65).

These facts prompted us to ask: Can the young guinea pig be sensitized as easily as the adult guinea pig? We sensitized new-born guinea pigs (2 days old and weighing between 60 and 80 gr.) by the epicutaneous application of a primary toxic solution of 1% DNCB in acetone, performed in the same manner as for adult guinea pigs (see p. 4). At the site of the sensitizing applications on the neck, the irritation recorded after 11 applications did not differ from that observed in adult guinea pigs (N. Hunziker and G. Schinas, 1961).

## Results

### a) Macroscopy

Two to three days following sensitization, we performed a patch test on the guinea pigs (now 15 to 16 days old) with a 0.1% solution of DNCB in acetone. After 14 hours, the patch test was clearly positive in all 11 guinea pigs; controls of the same age did not react to the same solution.

### b) Histology

We examined the flank and the nipple 4, 6, 8 and 14 hours after the patch test.

After a single application of 0.1% DNCB in acetone, the histological examination of the nipple and the flank of *unsensitized* new-born guinea pigs, does not reveal anything in particular.

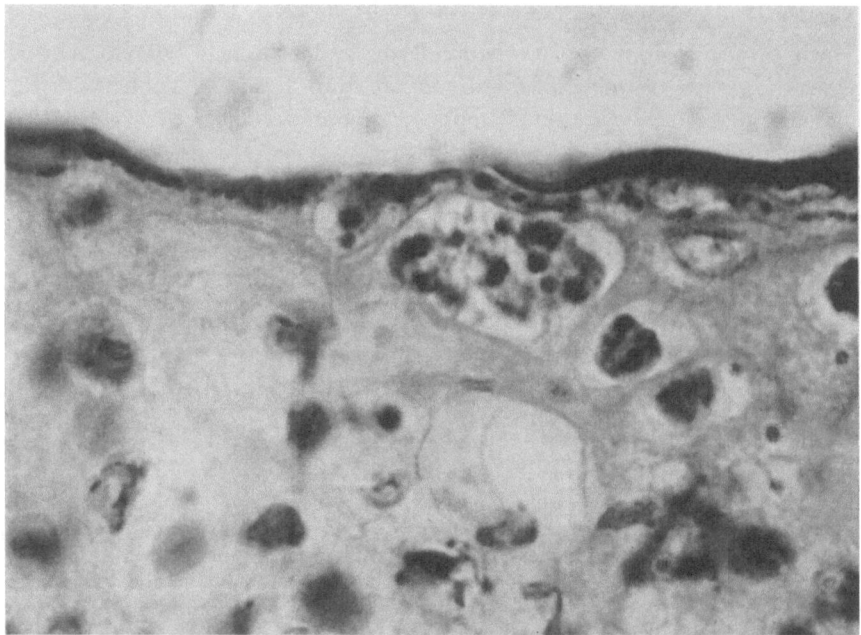

Fig. 15. Guinea pig 16 days old. Nipple 14 hours after a single application of 0.1% DNCB in acetone: subcorneal ,,vésiculette"

Histological examination of the nipple and flank of *sensitized* young guinea pigs distinctly reveals eczema. As in adult guinea pigs one can see a dermal infiltrate formed in part of small round cells, clear spongiosis of the epidermis and in addition, subcorneal "vésiculettes" (Unna, 1894; Civatte, 1925), in 8 out of 9 of these guinea pigs (Fig. 15) (Table 3).

Table 3. *Eczema due to dinitrochlorobenzene in the new born guinea pig*

| Guinea pig No. | No. of applications 1% on neck | Age of guinea pigs at beginning of sensitization | Age of guinea pigs at the time of tests | Epicutaneous test to 1‰ DNCB — Flank macroscopy[a] 14 h | Flank Histology[b] 4 h | 6 h | 8 h | 14 h | Nipple Histology[b] 4 h | 6 h | 8 h | 14 h |
|---|---|---|---|---|---|---|---|---|---|---|---|---|
| 7023 | 11 | 2 days | 16 days | + | | | | ++ | | | | ++V |
| 7024 | 11 | 2 days | 16 days | + | | | | ±+ | | | | +++V |
| 7025 | 11 | 2 days | 16 days | + | | | | ±++ | | | | +++V |
| 7289 | 11 | 2 days | 15 days | + | | − | | +++ | | | | ++V |
| 7290 | 11 | 2 days | 15 days | + | | ±V | | +++ | | + | | ++V |
| 7291 | 11 | 2 days | 15 days | + | | +V | | +++ | | +V | | ++ |
| 7292 | 11 | 2 days | 15 days | + | + | | ++ | ++ | | | + | |
| 7293 | 11 | 2 days | 15 days | + | ± | | ++ | ++V | | | + | |
| 7294 | 11 | 2 days | 15 days | + | − | | ++V | + | − | − | − | |
| 7334 | 0 | | 15 days | − | | | − | − | | | | − |
| 7335 | 0 | | 15 days | − | | − | − | − | | | − | − |
| 7336 | 0 | | 15 days | − | | − | − | − | | | − | − |
| 7026 | 0 | | 16 days | − | | | | − | | | | − |
| 7029 | 0 | | 1 day | − | | | | − | | | | − |
| 7030 | 0 | | 1 day | − | | | | − | | | | − |
| 7031 | 0 | | 1 day | − | | | | − | | | | − |

[a] Macroscopy: − no reaction, + homogeneous erythema. [b] Histology: − no eczema, ± outlines of spongiosis, + spongiosis, ++ widespread spongiosis, V subcorneal "vésiculette".

As previously stated (see p. 21), these subcorneal "vésiculettes" are very rarely observed in adult guinea pigs and then only in those strongly sensitized. We do not believe it to be a question of primary toxic effect, as (MIESCHER, 1952; CHARPY and coll., 1954; BANDMANN, 1960) reported for human beings, since we never found these "vésiculettes" in unsensitized new-born guinea pigs. In new-born guinea pigs, the subcorneal "vésiculettes" are doubtlessly a symptom of eczematous sensitization, as has been concluded for man: POLAK and MOM (1949), CIVATTE (1950), TZANCK and MELKI (1954). However, we cannot confirm whether or not this "vésiculette" is the "primordial vésiculette" of CIVATTE (1925), that is, a symptom of eczema which precedes spongiosis for, as shown in Table 3, the subcorneal "vésiculettes" found in the new-born guinea pigs were always accompanied by spongiosis.

*In conclusion, we can say that new-born guinea pigs can be rendered sensitive, like adults, to an eczematogenic substance (DNCB).*

As for the experimental eczema of guinea pigs, there is neither a paranatal "immunological immaturity" nor a particular immunity in the new-born guinea pigs.

This fact explains the virtually negative results in the experiments of ROSENTHAL and BAER (1963) and the negative results of SCHIMPF and FILIPP (1965) in their attempts to obtain immunotolerance (or allergotolerance) in adult guinea pigs, by a paranatal treatment with an allergen such as DNCB (see p. 68).

As for the form of the eczema, it should be pointed out that the subcorneal "vésiculettes" are frequent in new-born guinea pigs, but very rare in adult guinea pigs. This is a discrepancy distinguishing the eczema of the two ages. Otherwise, the histological examination of the new-born is similar to that of the adult.


# F. Appearance of Histological Eczematous Reactions During Sensitization

In order to study the appearance of histological lesions during sensitization, GOLAY and BRUN (1958) made 1, 2, 3, 4, 6, 9, 12, 15 or 18 daily applications of 1% DNCB in acetone on the neck of guinea pigs. After every application, they applied a solution of 0.1% DNCB in acetone on the nipple. This triggering application on the nipple was always performed at the same time as the last "sensitizing" application to the neck. The nipple was excised 24 hours later for histological examination.

We found that:

a) Two days after the beginning of the sensitization treatment, very slight spongiosis can already be seen and in one case even a well formed spongiosis is noted.

b) Five days after the beginning of the treatment, the eczematous lesions on the epidermis of the nipple are very marked. An infiltrate appears in the dermis. This infiltrate is essentially composed of histiocytes and not of lymphocytes.

c) Eight to 17 days after the start of sensitization, the spongiosis were very clear and the infiltrate had the same appearance as previously (Table 4).

An infiltrate composed of histiocytes has also been frequently noticed in human eczema (GANS and STEIGLEDER, 1955).

Table 4. *Appearance of histological eczematous reactions during sensitization (DNCB)*

| Guinea pig No. | No. days[a] | Histological examination of the nipple 24 h. after one application of DNCB 0.1% (in acetone) | |
| | | Epidermal reactions (spongiosis, exocytose) | Dermal reaction (edema, infiltrate) |
| --- | --- | --- | --- |
| 1 | 0 | — | — |
| 2 | | — | — |
| 3 | | — | — |
| 4 | | — | — |
| 5 | 1 | — | — |
| 6 | | — | — |
| 7 | | — | — |
| 8 | | — | — |
| 9 | 2 | + + | + |
| 10 | | — | — |
| 11 | | + | — |
| 12 | | ± | — |
| 13 | | ± | — |
| 14 | 3 | — | — |
| 15 | | + | — |
| 16 | 5 | + + | + |
| 17 | | + + + | + + |
| 18 | | + + | + |
| 19 | 8 | + + | + + |
| 20 | | + + + | + |
| 21 | | + + + | + |
| 22 | 11 | + + | + |
| 23 | | + + | + |
| 24 | | + + + | + + + |
| 25 | 14 | + + | + + + |
| 26 | | + + + | + |
| 27 | | + + | + |
| 28 | 17 | + + | + + |
| 29 | | + + | + |
| 30 | | + + + | + + |

[a] Number of days elapsed since the beginning of the sensitization treatment on the neck. This treatment was performed at the same time as the test on the nipple.

*In conclusion, the histological examination yielded an astonishingly short "incubation time" much shorter than that foreseen by the macroscopic examination:* FREY and WENK (1956) obtained macroscopic reactions only 6—9 days after the beginning of sensitization.

# G. Sensitization by Multiple Applications on the Nipple

In order to demonstrate a general sensitization, most of the experiments on the guinea pig that we have described up to now were performed on animals whose sensitization was effected at a determined place (the neck) and in whom triggering of the eczema (a single application) took place at a location never previously in contact with the eczematogenic substance (flank or nipple). However, contact eczema in man is most often due to repeated contact of the eczematogenic substance on the same site.

It would be interesting to know when the eczematous lesions appear at the location where the multiple sensitizing applications are performed. It would equally be interesting to know when the acanthosis appears and if there is a relation between the two phenomena (GOLAY and BRUN, 1958).

The 2 flanks of unsensitized guinea pigs are shaved. Every day a primary non-toxic acetonic solution of 0.1% DNCB, is applied both to the 2 flanks and the 2 nipples. A biopsy of the flank and of the nipple is performed after 3, 8, 10, 12 and 16 applications.

*After 3 applications*, there are no eczematous lesions, on the nipple, or on the flank and no acanthosis is observed.

*After 8 applications*, spongiosis appear on the epidermis of the flank and even more so on the epidermis of the nipple. A light dermal infiltrate of similar appearance is evident both on the flank and on the nipple; there is a clear acanthosis on the flank, but its appearance on the nipple is difficult to note in view of the intensity of the eczematous, epidermal lesions.

*After 10, 12 and 16 applications*, the eczematous lesions increase in number and intensity, as much on the flank as on the nipple. Acanthosis of the flank increases steadily.

The observation that the acanthosis appears late on the 8th day, at the same time as the first eczematous lesions, and that they increase in a parallel way, permits us to state that it is an "eczematous acanthosis".

We have seen elsewhere that if one applies daily a solution of 0.1% DNCB in acetone to a single nipple of a nonsensitized guinea pig (5, 8, 20 applications), acanthosis increases regularly on this nipple at the same time as the eczematous lesions are developing. Some eczematous lesions also appear and progressively grow in intensity on the flank and the other nipple, after a patch test (0.1% in acetone) (Table 5).

RAJKA and HARD (1960) applied a solution of 0.1% DNCB in alcohol 3 times a week for 1 year to the flank of a previously sensitized guinea pig, and they examined the treated places every 2 weeks. They observed that the acanthosis remained unaltered during this treatment. They also attributed this acanthosis to the allergic reaction triggered by the allergen (DNCB).

There are many authors who have concerned themselves, as we have done with the histology of the eczematous reaction of the guinea pig sensitized to DNCB. We will cite among others GINSBERG and coll. (1937), FREY and STUDER (1951), NILZEN (1952), ZELIGMAN (1954), FISHER and COOKE (1958), BAER and coll. (1956, 1957 and 1959), DE GRACIANSKY and coll. (1960), GRIMMER (1961), GRIMMER and SPIER (1961), MIESCHER (1962), NIEBAUER (1962), MACHER and SENNLAUB (1963), KLASCHKA (1964 and 1966).

Table 5. *Multiple applications of Dinitrochlorobenzene (DNCB) on the guinea pig's nipple*

| Number of daily applications of DNCB 0.1% (in acetone) on the left nipple | Number of guinea pigs (albinos, males, weight approx. 250 g) | Epicutaneous test 14 hours after a single application of 0.1% DNCB in acetone | | | | | | | | | | | | | |
|---|---|---|---|---|---|---|---|---|---|---|---|---|---|---|
| | | Macroscopy flank | | | Microscopy | | | | | | | | | |
| | | | | | *right nipple* not previously treated spongiosis | | | | *left nipple* previously treated (multiple applications) spongiosis | | | | | acanthosis |
| | | + | ± | − | ++ | + | ± | − | ++ | + | ± | − | | |
| 5 | 5 | 0 | 0 | 5 | 0 | 2 | 1 | 2 | 0 | 1 | 1 | 3 | | very light |
| 8 | 10 | 5 | 0 | 5 | 5 | 2 | 2 | 1 | 3 | 5 | 1 | 1 | | variable |
| 20 | 23 | 21 | 2 | 0 | 13 | 6 | 3 | 1 | 11 | 9 | 2 | 1 | | clear |

macroscopy
− no reaction
± doubtful reaction, a few disseminated erythematous locations
+ homogeneous clear erythema

histology
− no spongiosis
± outlines of spongiosis
+ spongiosis
++ widespread spongiosis

# II. Experimentation with Other Substances

## A. Sensitization with Paranitrosodimethylaniline (PNDMA)*

$$H_3C \diagdown \diagup CH_3$$
$$N$$

$$N=O$$

In 1935, LANDSTEINER and JACOBS sensitized guinea pigs by repeated injections of small doses of PNDMA. The tests to control the sensitization were made by an intra-dermal injection. These authors also report (without giving details) that sensitization could be obtained with PNDMA "by rubbing guinea pigs with vaseline containing 5% of the compound".

PNDMA is also capable of sensitizing man: LANDSTEINER, ROSTENBERG and SULZ-BERGER (1939), W. EPSTEIN and KLIGMAN (1958). Indeed, man can be sensitized with PNDMA in 36.6% of the cases (experiments in 82 individuals), whereas, with DNCB man can be sensitized in 50% of the cases. This shows that PNDMA would be a weaker sensitizing agent than DNCB (LANDSTEINER, ROSTENBERG and SULZBERGER, 1939).

### 1. Method of Sensitization

Sensitization was carried out by 14 applications of 5% PNDMA in vaseline on the shaved neck of the guinea pig.

### 2. Primary Toxic Reaction

One application of 5% PNDMA in vaseline to the neck of unsensitized guinea pigs produces nothing unusual; but if the number of applications is increased, lesions appear: redness, squames and crusts. Under histological examination, we observe epidermal changes (extracellular edema, vesicles, necrosis) and an infiltrate which is composed of lymphocytes and leucocytes.

One application of 1% PNDMA in olive oil produces no visible change on the flank or on the nipple under both macroscopic and histological examination after 3, 6 or 14 hours.

### 3. Specific Reaction due to Sensitization

#### a) Macroscopic Reaction

In sensitized guinea pigs, the macroscopic reaction which follows the patch test (1% PNDMA in olive oil) begins to appear within 2 to 3 hours. It is very distinct after 14 hours, it decreases after 24 hours and shades off progressively. We do not observe an erythema which is as confined as that observed with DNCB, because the excipient is different: olive oil for PNDMA, acetone for DNCB. This erythematous lesion is accompanied by an edema.

---

* PNDMA corresponds to PDMA as printed in certain American publications.

*b)  Histological Examination*

α) Flank

In the process of examining slides made shortly following the test we observed that histological lesions are already noticeable *after 2 hoours*: outlines of spongiosis, moderate infiltrate.

*After 6 hours:* one can already observe more or less extensive epidermal alterations, a moderate infiltrate and a distinct edema of the dermis.

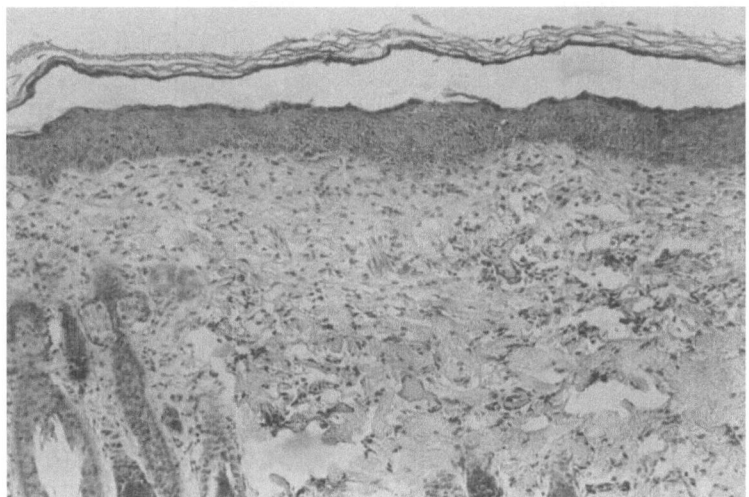

Fig. 16. Guinea pig sensitized to PNDMA. Flank 14 hours after a single application of 1% PNDMA in olive oil: acanthosis, no epidermal lesions

*After 14 hours:* in the first group of guinea pigs we observed a distinct acanthosis, which is not observed in unsensitized animals. Thus, this is not a matter of an intrinsic acanthotic effect of the substance, but it is due to the fact that the animal is sensitized (Fig. 16) (Table 6). We have never observed such an acanthosis after a patch test in guinea pigs sensitized to DNCB.

In a second group of sensitized guinea pigs, some also exhibited this acanthosis. On the contrary, in the others which did not present an acanthosis of the flank, we found small but distinct spongiosis (Fig. 17).

In the two groups, the infiltrate is not very extensive. It varies in composition, being composed of immature cells of connective tissue, small lymphocytes and occasionally some eosinophils (N. HUNZIKER and coll., 1964).

In a third group of guinea pigs who were sensitized in the same manner, we observed histologically only small changes in the epidermis. However, we see an infiltrate formed mainly by eosinophils with only some lymphocytes.

*After 24 and 48 hours:* we usually observed a moderate acanthosis at the level of the epidermal alterations; the infiltrate is moderate.

β) Nipple

*After 14 hours:* there are well formed spongiosis, occasionally intra-epidermal vesicles and an infiltrate which is rather moderate and composed of immature cells of connective tissue, small lymphocytes and eosinophils at times.

Table 6. *Sensitization of the guinea pig with Paranitrosodimethylaniline*

| Guinea pig No. | Number of applications of 5% PNDMA on the neck | PNDMA tests. Examination performed 14 hours after application | | | | | |
| --- | --- | --- | --- | --- | --- | --- | --- |
| | | Macroscopic results on the flank[a] | Histology of the flank[b] | | Histology of the nipple[b] | | |
| | | | spongiosis | acanthosis[c] | spongiosis | acanthosis[c] | |
| 945 | 0 | — | — | 9.8 | — | 16.6 | |
| 946 | 0 | — | — | 8.9 | — | 22.1 | |
| 947 | 0 | — | — | 10.9 | — | 23.1 | |
| 948 | 0 | — | — | 13.1 | — | 20.7 | |
| 53 | 0 | — | — | 11.9 | ± | 14.7 | |
| 54 | 0 | — | — | 9.3 | — | 25.2 | |
| 830 | 14 | + | ± | 20.9 | + | | |
| 831 | 14 | + | ± | 23.5 | + | 19.8 | |
| 836 | 14 | + | ± | 19.9 | ++ | 18.5 | |
| 837 | 14 | + | ± | 21.3 | + | 23.1 | |
| 838 | 14 | + | — | 18.7 | + | 19.4 | |
| 839 | 14 | + | + | 27.3 | ± | 22 | |
| 943 | 14 | + | ± | 16 | ++ | | |
| 944 | 14 | + | ± | 19.3 | + | 20 | |

a macroscopy: — no reaction
+ homogeneous clear erythema

b microscopy: — no spongiosis
± outlines of spongiosis
+ spongiosis
++ Widespread spongiosis

c The average thickness of the normal epidermis of the flank is 10, and that of the nipple 20.

We observe spongiosis on the nipple even if we do not observe it on the flank of the same guinea pig. It is surprising that we did not find similar lesions in both the flank and nipple. This shows that the lesions do not always evolve in the same manner on the flank and nipple (N. HUNZIKER, 1962).

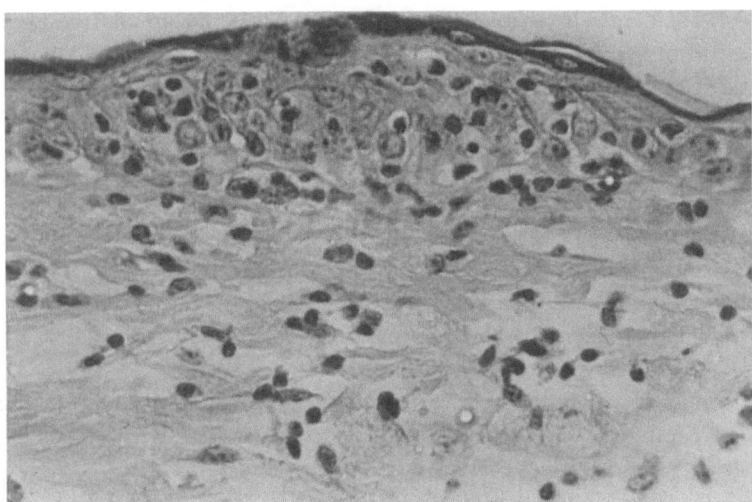

Fig. 17. Guinea pig sensitized to PNDMA. Flank 14 hours after a single application of 1% PNDMA in olive oil: no acanthosis, little spongiosis

To summarize: *in guinea pigs sensitized to PNDMA, the macroscopic test is generally distinct; whereas, histological examination yields a great variability in the epidermal as well as in the dermal lesions.*

# B. Sensitization with Citraconic Anhydride (CA)

$$CH_3-C-C\underset{HC-C}{\overset{O}{\diagup}}\diagdown O$$

In 1940, JACOBS, JACOBS and coll. described experiments with different anhydrides, one of which was citraconic anhydride. Their attention had been attracted by a case of asthma in a chemist described by KERN (1938/39); the asthma, however, was not due to citraconic anhydride but to phthalic anhydride.

Citraconic anhydride provokes specific reactions in sensitized guinea pigs (JACOBS; JACOBS and coll., 1940): an immediate urticarial reaction after 30 minutes, then a delayed reaction after 6 to 8 hours which subsequently disappears in a few days. The immediate reactions were occasionally accompanied by a generalized rash.

Sensitivity of guinea pigs to CA was also observed by CHASE (1954); however, no histological examination was performed.

## 1. Method of Sensitization

We applied pure CA to the neck of guinea pigs (11 applications in 12 days) and made 4 sub-cutaneous injections (0.05 ml) of a 1% solution of CA in olive oil.

At the end of the sensitization treatment, the neck is extremely irritated and the histological examination shows an extensive infiltrate composed largely of eosinophils.

## 2. Primary Toxic Reaction

1. If pure CA is applied to the nipple or to the flank of an unsensitized guinea pig, an erythema and edema appear at the site of the test. After 14 hours the histological examination shows epidermal lesions (vesicles, detachement of the epidermis, epidermal necrosis) and a dermal infiltrate which is partially composed of eosinophils which invade the epidermis.

If the conditions of the test were unknown (sensitization, concentration of the allergen used for the test), it would be difficult to distinguish these lesions from those of the specific reaction of sensitization (see below and p. 33).

2. If 25% CA (in dioxane) is applied to the nipple or to the flank of unsensitized guinea pigs, nothing unusual is observed, either under macroscopic or histological examination.

However, if colchicine is injected 24 hours after the test and 9 hours before the excision (a total of 33 hours after the test) we observe *an increase in mitotic activity* as with DNCB (p. 11).

## 3. Specific Reaction due to Sensitization

Two weeks after the end of sensitization, we performed as did JACOBS (1940), a scarification on the flank and we applied a 25% solution of CA (in dioxane). In addition, by applications of the solution, we performed a patch test on the other flank.

Ten to thirty minutes after the scratch-test and even after the patch test, an urticarial papule appears. This urticarial lesion appears in a much shorter time if the guinea pig is strongly sensitized. When the immediate urticarial reaction is at its maximum, there is occasionally a generalization of the reaction in the form of papules and there is sometimes even an erythrodermia. This reminds one of the generalized toxicodermal reactions in man (JACOBS, 1940). After about 7 hours, the urticarial lesion at the site of the test undergoes a progressive transformation into a delayed type of reaction, which is well defined, less papular, and less red. Unlike JACOBS (1940), we were unable to determine the moment at which the immediate reaction disappears and the delayed reaction appears.

However, we must report that in sensitized guinea pigs we have not always observed an immediate urticarial reaction; whereas, a delayed reaction appearing 7 hours after the test, has always been present, with one exception.

*Histological Examination*

α) Flank

Even 30 minutes after the test, but especially after 2 hours, when the epidermis does not show distinct alterations, we find intravascular and perivascular eosinophils in the dermis (Fig. 18).

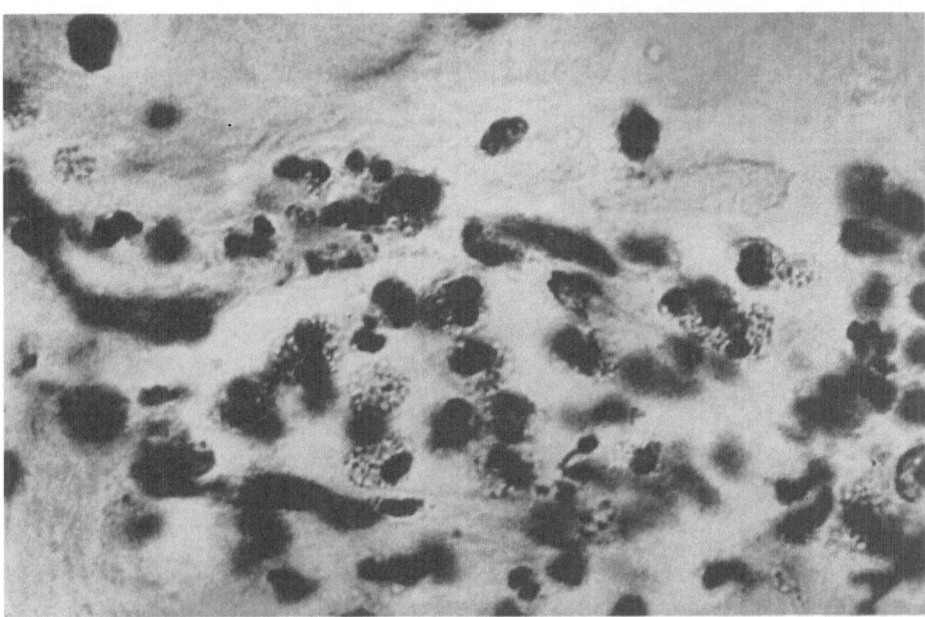

Fig. 18. Guinea pig sensitized to CA. Flank 1 hour after a single application of 25% CA in dioxane; immersion: intravascular and perivascular eosionophils

After 7—24 hours the leucocytic infiltrate increases and is predominantly composed of eosinophils which often invade the entire epidermis. There is a slight acanthosis, especially after 24 hours, some spongiosis but not in a regular manner. The alkaline phosphatase reaction reveals some lymphocytes both in the dermis and epidermis, but this reaction is less pronounced when compared to that in eczema due to DNCB.

β) Nipple

Fourteen hours after the test, we observe spongiosis and intra-epidermic vesicles, as seen in the eczema caused by DNCB. However, the dermal infiltration is much more extensive with citraconic anhydride and contains lymphocytes and above all numerous eosinophils which invade the epidermis (Fig. 19). In eczema due to DNCB the infiltration is mainly by lymphocytes.

The phosphatase alkaline reaction is much weaker than in the eczema due to DNCB. It seems that as the number of eosinophils in the infiltrate increases, the number of lymphocytes decreases.

3  Hunziker, Experimental Studies

It is not surprising to find intra- and perivascular eosinophils in the urticarial phase, as leucocytic infiltration with many eosinophils is characteristic of urticarial reactions in man (BERGER and LANG, 1932; W. JADASSOHN, 1932). However it is surprising that so many eosinophils* (14 hours and more after the test) are seen at the same time as the eczematous epidermic lesions (spongiosis and intraepidermic vesicles).

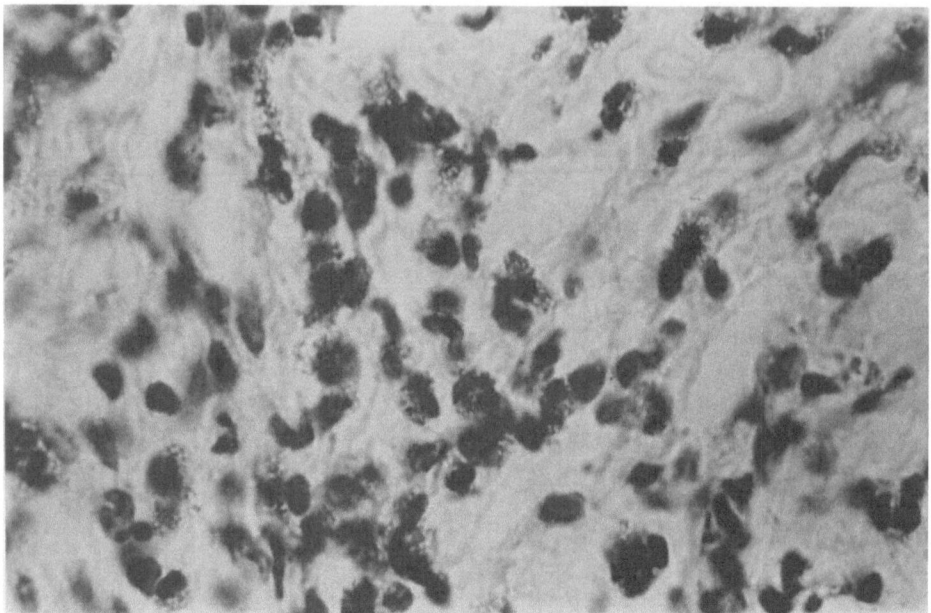

Fig. 19. Guinea pig sensitized to CA. Nipple 14 hours after a single application of 25% CA in dioxane; immersion: numerous eosinophils

MUSSO and BRUN (1962) have shown that in the guinea pig sensitized with "egg white", the infiltrate is composed of eosinophils.

To summarize: *in guinea pigs sensitized to CA, we see an eosinophilic infiltrate which persists for 24 hours or more; this delayed infiltrate can be accompanied by eczematous lesions and is preceded by urticarial lesions* (N. HUNZIKER, 1964).

# C. Sensitization with Propionic Anhydride (PA)

$$CH_3-CH_2-C\diagup^O_{\diagdown O} \diagdown_{\diagup} CH_3-CH_2-C\diagdown^O_{\diagdown O}$$

JACOBS (1940) ends a study on the immediate reaction in guinea pigs with the following: "Excellent sensitization-effects have been produced by the same method

* Guinea pigs sensitized with AC show an blood eosinophilia (p. 45).

with a third substance, propionic anhydride, using patch tests with a 25% dioxane solution. Scratch-tests, however, did not yield definite wheal-and-erythema reactions . . .".

JACOBS's observations on PA are limited to this remark. For this reason we thought it interesting to expand the research on the sensitization of guinea pigs with PA.

## 1. Method of Sensitization

Sensitization was carried out by 11 applications in 12 days of pure PA to the neck and by 4 intradermal injections (0.05 ml of a 1% solution of PA in olive oil).

## 2. Primary Toxic Reaction

1) If pure PA is applied to the flank and nipple of unsensitized guinea pigs, a light redness is observed after 14 hours. Under histological examination the lesions of the flank and nipple are numerous: intra-epidermal vesicles, epidermal detachment, epidermal necrosis, dermal infiltrate with numerous eosinophils which also invade the epidermis.

2) When a 25% solution of PA (in dioxane) is applied to the flank (16 guinea pigs) or to the nipple (6 guinea pigs), no lesions are observed after 14 hours either macroscopically or histologically.

If 24 hours after the application, a colchicine injection is administered (blockage of mitosis, page 11 and 32) an increase in mitotic activity is not observed.

## 3. Specific Reaction of Sensitization

Two weeks after the end of sensitization we performed a scratch test on one flank and a patch test, by application of 25% PA in dioxane, on the other. After 20—30 minutes, we did not observe an urticarial reaction, as we had in the sensitization with citraconic anhydride.

Approximately 2 hours after the application, a limited redness and a slight edema appear. In the hours that follow, the redness and degree of infiltration increase. The intensity of the reaction increases up to 24 hours. After 48 hours the reaction is still very intense and the epidermis begins to desquamate; then the reaction disappears little by little in the days which follow.

### Histological Examination

α) Flank

The histological results are similar after scarification and a simple epicutaneous application.

*After 1 hour*, there are no epidermal lesions, except for some swollen cells and occasionally a disordered basal layer. In the dermis, there is an infiltrate which contains a few eosinophils.

*After 14 hours*, the epidermal lesions are significant: in certain places the epidermal cells are extremely swollen (pseudo-acanthosis*), elsewhere we see an atrophy of the epidermis and next to it, either a pseudo-acanthosis, or an acanthosis. Intra-epidermal vesicles are frequently seen (Fig. 20).

---

* Pseudo-acanthosis = thickening of the epidermis not due to the increase of cellular layers but to their swelling.

3*

In some sections more or less extended *epidermal necrosis* and detachment of the epidermis are seen, in others intra-epidermal abcesses may be observed. In the upper dermis, the small vessels are dilated and there are extravasations. The infiltrate is very extensive, being composed of small round cells, connective tissue cells, and eosinophils. The eosinophils are intra- and perivascular. We cannot say, as we did in the sensitization with CA, that the infiltrate is largely composed of eosinophils, but we can affirm that these are much more numerous than in the sensitization with DNCB.

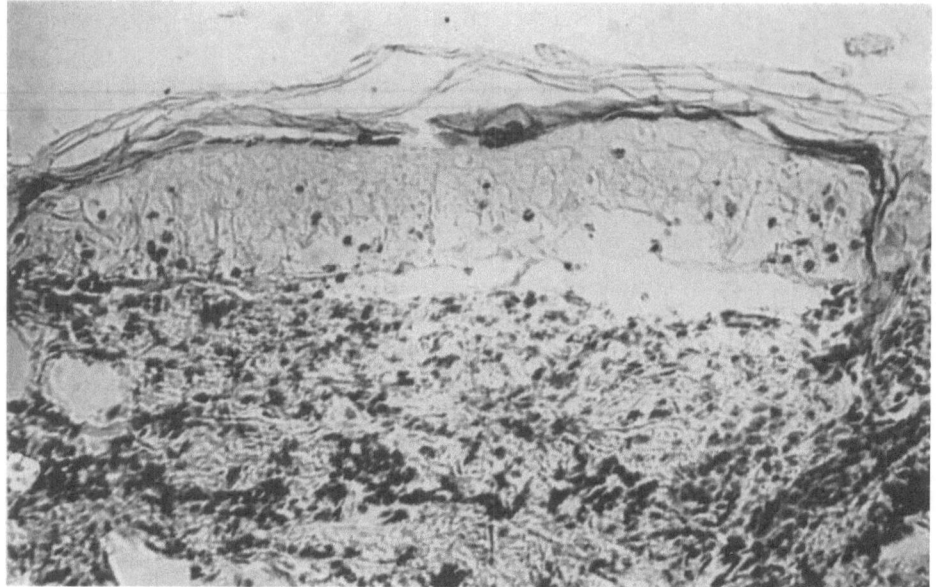

Fig. 20. Guinea pig sensitized to PA. Flank 14 hours after a single application of 25% PA in dioxane: epidermal necrosis, infiltrate

The reaction with alkaline phosphatase reveals a great number of positive cells in the infiltrate. We also note black points invading the epidermis.

*After* 24 *hours*, the epidermal lesions are less intense, even if the macroscopic reaction is still very pronounced. There is an acanthosis. The infiltrate is still very extensive and the extravasations still very marked.

β) Nipple

*After* 14 *hours*, contrary to what we observe on the flank, the epidermal lesions on the nipple are not very extensive. We see swollen cells and occasionally spongiosis or outlines of spongiosis. Exceptionally the epidermis is completely invaded by the cells of the dermal infiltrate. The infiltrate is more or less extensive and contains some eosinophils. The reaction with alkaline phosphatase shows positive cells in the infiltrate.

To summarize, we are surprised by the histological difference between the response to the patch test on the flank and on the nipple. Usually the lesions are better formed and more easily visible on the nipple, where the epidermis is thicker, as we noted in the sensitization with DNCB, and even with CA. The results are just the opposite with PA; there are more lesions on the flank than on the nipple (HUN-ZIKER and coll., 1965).*

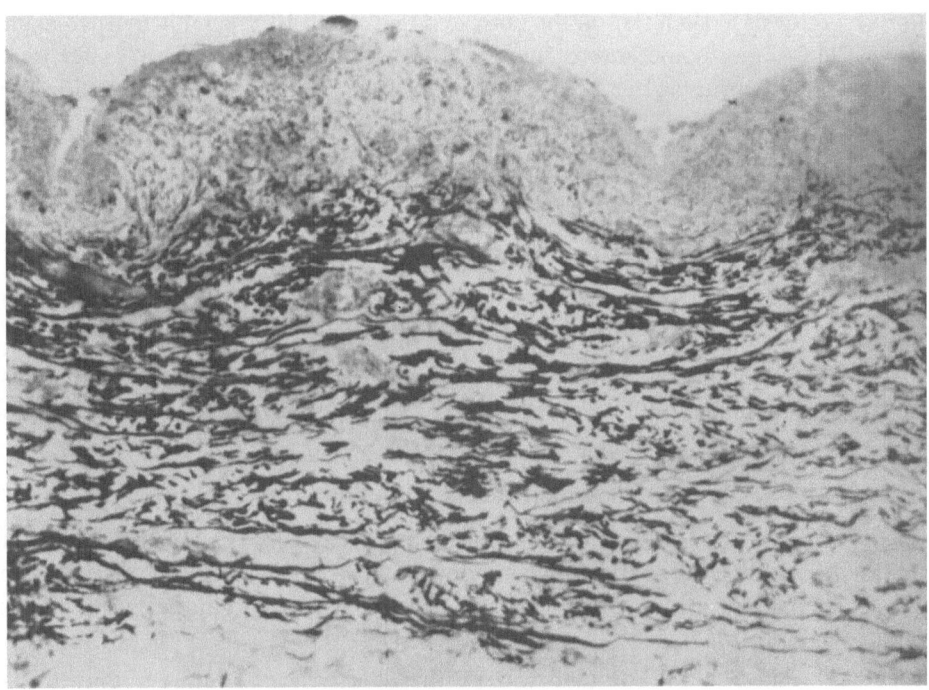

Fig. 21. Toxic epidermal necrolysis in man (Lyell's Disease)

The extensive lesions that we have observed on the flank 14 hours after the tests, to a certain extent bring to mind a skin disease in man, which is so serious that it can lead to the death of the patient:

*Lyell's diseases* (toxic epidermal necrolysis) (see p. 49 and 69) (cf. SOLTERMANN, 1959; ZAK and coll., 1964 and others) (Fig. 21).

# D. Sensitization with Paraphenylenediamine

$$NH_2$$

$$NH_2$$

In man we know of the contact eczema due to azo dyes (see among others MAYER, 1928 and 1929; DOBKEVITCH and BAER, 1947).

---

* A difference in the response of the flank and nipple to X-rays has also been reported by MAGGIORA, BUJARD and W. JADASSOHN (1965).

This eczema is the consequence of a sensitization by paraphenylenediamine or one of its derivatives. It is not, however, frequent when one thinks of the number of people who have their hair dyed!

In 1931, MAYER sensitized the flank of guinea pigs by applications of 10% paraphenylenediamine in vaseline. Usually 4 weeks after the beginning of sensitization he performs a patch test on the untreated flank with 10% paraphenylenediamine in vaseline (under some sort of jacket) („Festschließende Jäckchen"). The tests were distinctly positive after 12 and 24 hours on the guinea pigs whose sensitization was successful.

## 1. Method of Sensitization

We sensitized 12 guinea pigs on the neck with 11 applications (within 12 days) of 20% paraphenylenediamine in vaseline.

## 2. Primary Toxic Reaction

We did not observe a primary toxic reaction either under macroscopic or under histological examination, after applying 10% paraphenylenediamine in vaseline on the flank or on the nipple.

## 3. Specific Reaction of Sensitization

In order to estimate the sensitization ,we applied 10% paraphenylenediamine in vaseline on the shaved flank but without the "jacket". We did not observe, after 12—24 hours, any redness or infiltration in the 12 sensitized guinea pigs (N. HUN-ZIKER and coll., 1964).

MAYER and SULZBERGER (1931) write that it is easier to sensitize guinea pigs with paraphenylenediamine when they take their winter food. In our animals, sensitization was carried out during winter and in spite of that, they were not sensitized.

### Histological Examination

α) Flank

After one application of 10% paraphenylenediamine in vaseline, we observed no histological reaction.

β) Nipple

We examined the nipples after one application of 10% paraphenylenediamine in vaseline. We did not note any histological lesions on any of the slides, except on the slide of one guinea pig which showed clear spongiosis and a slight infiltrate composed partly of lymphocytes (Fig. 22).

To summarize, there was only one guinea pig out of 9, whose nipples we examined, which had been sensitized, and even this sensitization is apparent only under histological examination*. We could say that this was an "idiosyncratic", guinea pig, that is to say quite easier "sensitizable" than others.

---

* MACHER and SENNLAUB (1963) also observed on 2 guinea pigs sensitized with DNCB typical histological pictures of eczema, whereas the macroscopic examination was completely negative.

MAYER (1931) was surprised by the histological resemblance between the patch tests (after 24 hours) in guinea pigs sensitized with paraphenylenediamine and the histological picture in human eczema: spongiosis, "altérations cavitaires", intra-epidermal bullae, small pustules filled with polynuclear cells, infiltrate of the superficial dermis composed of polynuclear cells and invading the epidermis. Paraphenylenediamine induced an acanthosis on the flank which, according to MAYER, favors the appearance of spongiosis. But is this acanthosis due to vaseline and not to paraphenylenediamine itself or to sensitization?

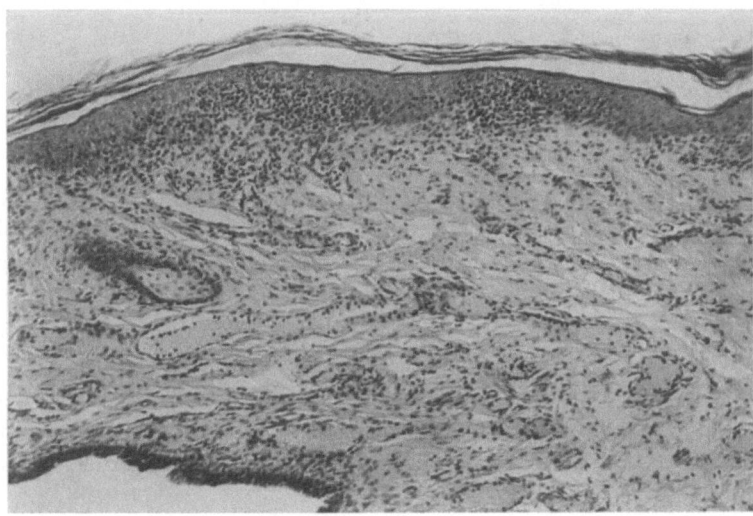

Fig. 22. Guinea pig sensitized to paraphenylenediamine. Nipple 14 hours after a single application of 10% paraphenylenediamine in vaseline: spongiosis. Macroscopic examination showed no erythema on the flank

CORDONNIER (1949) studied the eczema of the guinea pig caused by paraphenylenediamine using two methods: Feulgen's reaction for thymonucleic acid and Brachet's reaction for thymo- and ribonucleic acids. On an eczematous epidermis, both staining methods show that the nuclear degeneration is accompanied by the persistence of high contents of DNA up to the most superficial layers of the epidermis. This is the opposite of what is seen in a normal epidermis where the content of DNA in the chromatin diminishes nearer to the surface. When a biopsy is performed on a previously eczematous site where the clinical signs have disappeared, the perturbations of the nucleic acids of the nuclei are found to persist for a long time.

Therefore, the author has shown a perturbation of nucleic acids which continues after the healing of the histological lesions of eczema.

In 1951 and 1954, DUESBERG sensitized guinea pigs with paraphenylenediamine (10% in vaseline) by 8 daily applications and tested the animals on the 20th day after the beginning of sensitization, by an intradermal injection of a 2% aqueous solution.

Histologically, he observes zones of hyperplasia alternating with zones of atrophy, but the latter are rather predominant. There is hyperkeratosis, and at places ulceration of the epidermis. In only one slide there were a few locations of spongiosis. In the dermis, the inflammation is very discrete; there is a predominance of necrotic phenomena, with at places a dissociation of the dermal fibres. The vascularisation is not very significant.

ZELIGMAN (1957) made several applications of vaseline containing paraphenylenediamine (2%) on the flank: on the 10th application 3 out of 4 guinea pigs show a macroscopic reaction. Under histological examination: there is an acanthosis*, exocytosis, eosino-phils and some polynuclear leucocytes. The histological preparations of the 4th guinea pig, which has a negative macroscopic reaction, show, however, some modifications: acanthosis*, slight exocytosis, moderate infiltrate.

We wanted to cite the experiments of those who have shown that it is possible to obtain a sensitization with paraphenylenediamine, whereas, our results were negative with one exception. This exception shows that the macroscopic examination is not sufficient to ascertain a failure of sensitization.

## E. Experiments Concerning Sensitization to Nickel

Sensitization to nickel is of practical as well as theoretical interest.

Nickel is also known as an eczema -producing substance, especially in women (suspenders, hooks, earrings, wristwatches etc.). CALNAN (1957) even spoke of an "epidemic" of eczema due to nickel in London. Such incidents seem to have dimi-nished since the widespread use of plastic in place of nickel.

In contrast to the frequency of eczema caused by nickel in women and to certain remarks in the literature, there are, at least in Switzerland, few workers who are sensitized to nickel**. In the nickel workshops of Geneva, we have noticed that some young workers quickly become sensitive and must stop work during their apprentice-ship, while others, fortunately the majority, can pursue their trade without harm (N. HUNZIKER, MUSSO, 1959).

In 1926 WALTHARD had already attempted to sensitize guinea pigs to nickel.
COCA and MILFORD (1934) were not able to confirm this sensitization.
STEWARD and CORMIA (1934) induced a dermatitis to nickel in guinea pigs by repeated applications of a nickel solution; but according to these authors, this was a phenomenon of irritation and not of specific hypersensitization.
GRAUL and KALKOFF (1948/49) claim to have succeeded in sensitizing guinea pigs to nickel.
NILZEN and coll. (1955) sensitized 1 out of 5 guinea pigs with epicutaneous applications of a nickel sulfate solution to which they added a surface active agent: sodium lauryl sulfate. The eczema was distinct under macroscopic and histological examination.
VINSON and CHOMAN (1960) also report positive experiments in the sensitization of guinea pigs to nickel.

### Methods and Results of Sensitization

a) We also attempted to sensitize guinea pigs by epicutaneous applications to the flank of a 5% nickel sulfate solution with either lauryl sulfate (according to NILZEN and WIKSTRÖM, 1955), or Triton X 100 as surface active agents.

We have not been able to demonstrate sensitization with our animals.

b) Within an interval of one week, we injected the neck twice with a nickel solution containing Freund's adjuvant (2 mg $NiSO_4$ per injection) without pro-voking sensitization. This is unlike the sensitization provoked by potassium bi-chromate (see p. 41).

---

* But, paraphenylenediamine in vaseline was applied several times and we know that vaseline can produce an acanthosis.
** W. JADASSOHN and SCHAAF (1928/29).

c) After having irritated the skin of the flank twice a day for three days with black soap, we applied a 30% Nickel sulfate solution to which Triton X 100 at 1% had been added without obtaining sensitization.

Indeed, we had thought that preliminary irritation favored sensitization, since BURCKHARDT (1935) had shown with man that, after preliminary treatment with alkali, he could obtain sensitization to nickel. However W. JADASSOHN and SCHAAF (1928/29) had not been able to induce sensitization in man by repeated applications of a nickel sulfate solution but without alkali.

In 1963, VANDENBERG and W. L. EPSTEIN were able to sensitize man to nickel by the following method: irritations, occlusive bandages and repeated exposures of the same region. They found that prolonged exposure to the allergen increased the frequency of sensitization. There was no cross-sensitivity with other metals.

d) We then tried to follow WALTHARD's technique (1926) as closely as possible: application 3 times a day of a 1% nickel sulfate solution to the left buttock of the guinea pig, until the appearance of an inflammatory dermatitis. Subsequently, the same treatment is given to the right side. But, contrary to the observations of WALT-HARD, we did not note any significant difference between the two sides in time of appearance of the lesions. We did not obtain sensitization.

e) Finally, we injected a suspension of Raney nickel (finely divided nickel used as a catalyst in certain organic chemistry hydrogenations). This suspension provoked a granuloma at the site of injection, but did not result in sensitization.

*To detect sensitization*, we performed a patch test with a 5% nickel sulfate solution to which 1% Triton X 100 was added on all the animals sensitized by these different methods. We also made intradermal injections with 0.01% to 0.1% nickel chloride. ST. EPSTEIN (1956, 1958) has shown in man that with certain eczema producing substances (nickel, chromate, penicillin, rivanol, gentian violet, neomycin), tests by intradermal injection can be positive, even when patch tests are negative. These tests were carried out 14 days after the end of the sensitization period and we repeated them from week to week, sometimes for as long as 80 days. All the tests were negative\*. *Thus, we did not succeed in sensitizing the guinea pig by any of the preceding methods.*

Moreover, the seasonal factor is not significant since these experiments were spread over an entire year.

These negative experiments with nickel show that, if there is any similarity between the sensitization of the guinea pig and of man (primine, DNCB, phenyl-hydrazine), this resemblance is sometimes difficult to demonstrate (N. HUNZIKER, 1960).

# F. Sensitization with Hexavalent Chromium

Eczema caused by cement is the most widespread occupational dermatosis in Switzerland as well as in other countries (Musso and coll., 1962). The association between sensitivity to cement and to potassium bichromate, is universally acknowledged. In masons suffering from eczema, patch tests with cement were less often positive, than those with 0.5% potassium bichromate. BRUN (1963 and 1964) in our department,

---

\* Unfortunately we have not done histological examinations.

succeeded in making these tests with cement positive much more often by adding sodium sulfate to the diluted cement to elute the adsorbed chromium.

NILZEN and WIKSTRÖM (1955) were successful in sensitizing 5 guinea pigs to hexavalent chromium by the application of a mixture of 0.5% potassium bichromate plus a wetting agent, 1% lauryl sodium sulfate. Such a sensitization had not been possible without the wetting agent. The patch tests with 0.5% potassium bichromate mixed with 1% lauryl sulfate were positive, in the macroscopic as well as in the histological examinations: epidermis thicker than normal, intra- and extra-cellular edema, leucocytes in the epidermis, dilation of vessels in the superficial layer of the dermis and invasion of leucocytes especially at the junction of the epidermis and dermis.

MAYER and coll. (1958) also sensitized guinea pigs to bichromate in nearly the same manner as NILZEN and coll. (1955), but they demonstrate sensitization by intradermal injections of 0.0025% potassium bichromate.

An exaggerated reaction, considered as a hypersensitivity of guinea pigs to chromium, has also been described by WIKSTRÖM (1962). (This hypersensitivity of the skin has been transmitted by parabiosis to only 1 animal in 12 experiments.)

Other experiments of sensitization with the epicutaneous application of hexavalent and trivalent chromium are described in DA FONSECA's book (1963), and by SCHWARZ-SPECK and KEIL (1965).

## 1. Method of Sensitization

a) For 8 days we applied to the neck of the guinea pig a 0.5% solution of potassium bichromate containing a wetting agent (Triton X 100).

b) Using CHASE's technique, we injected twice in the neck a potassium bichromate preparation to which Freund's adjuvant was added (0.5 gr potassium bichromate/ml). These 2 injections were separated by an interval of one week.

## 2. Results

a) In sensitizing by 8 *applications* of a potassium bichromate solution containing a wetting agent, we did not obtain a clearly positive patch test (4 were questionable, 5 were negative). We continued the application (7 and 14 applications), but none of the patch tests became unmistakably positive. We made no histological examination.

b) Twenty to 40 days after the beginning of the sensitizing injections, we performed a test by epicutaneous application to the flank and nipple of a 0.5% potassium bichromate solution containing 1% Triton X 100. We also tested the flank by intradermal injection of 0.0025% potassium bichromate according to MAYER and coll. (1958).

aa) Macroscopically, the patch tests on the flank gave questionable or negative reactions, there was rarely a positive reaction.

Histological examination of these reactions only rarely showed true spongiosis and, in 2 out of 14 animals, infiltrates in the dermis were observed.

Similarly, there were few if any histological lesions on the nipple. These histological examinations were made 14 hours after one application of 0.5% potassium bichromate.

bb) Tests by intradermal injection of 0.0025% potassium bichromate regularly gave a very distinct reaction, already discernible after 14 hours and growing more marked during the next 24 to 48 hours. The controls (non-sensitized guinea pigs) gave no reaction.

Histological examination of this reaction shows a significant infiltrate in the dermis composed of polynuclear leucocytes and lymphocytes. There is also a very pronounced edema. Well defined spongiosis is not seen.

Thus, histological examination reveals that the reaction of the dermis is much stronger than that of the epidermis. This corresponds to the observations made in man by St. Epstein (1956 and 1958) with nickel, chromium, neomycin, penicillin, rivanol, and gentian violet. It seems that the epidermis and dermis do not always respond simultaneously to the provocation of the allergen (see p. 70).

c) On 6 guinea pigs whose sensitization to potassium bichromate (hexavalent chromium) had been proven by intradermal injection, we performed tests with trivalent chromate (basic sulfate of Cr III, prepared according to O. Flaherty, see Morris, 1958). First, we injected this trivalent chromium into 5 control guinea pigs and observed only negative tests. The intradermal injection of trivalent chromium into guinea pigs sensitized to hexavalent chromium did not trigger a reaction, not even a dubious one. Under histological examination, we noted very little infiltrate in the dermis. In the epidermis we occasionally noticed minute spongiosis.

Hence, cross-sensitivity between trivalent and hexavalent chromium was not convincingly manifested.

However, Schwarz-Speck and Keil (1965), as well as Da Fonseca (1963), have recorded a cross sensitivity between hexavalent and trivalent chromium in guinea pigs.

# G. Sensitization with Two Eczematogens

## 1. Dinitrochlorobenzene (DNCB) Followed by Paranitrosodimethylaniline (PNDMA)

We initially sensitized 12 guinea pigs with DNCB (technique p. 4), verifying the sensitization after 14 days by a patch test which was strongly positive in all the animals. We subsequently sensitized them with PNDMA (technique p. 28).

Two weeks after this second sensitization, we tested one flank with 0.1% DNCB in acetone and the other with 1% PNDMA in olive oil. The 2 patch tests were strongly positive (N. Hunziker and coll., 1964). Thus, we can say that previous sensitization with a strong eczematogen, such as DNCB, has not prevented further sensitization with a weaker eczematogen, as Landsteiner, Rostenberg and Sulzberger (1939) have also stated.

Our experiments can be compared with similar experiments in man.

W. L. Epstein and Kligman (1958), using the same 2 eczematogens in man, showed that sensitization with a strong eczematogen, such as DNCB, diminishes man's capacity to be sensitized with a weaker eczematogen, such as PNDMA.

## 2. Paranitrosodimethylaniline (PNDMA) Followed by Dinitrochlorobenzene (DNCB)

We treated 18 guinea pigs to PNDMA (technique p. 28). Nine animals were strongly sensitized (positive patch test) and 9 were weakly or not at all sensitized, despite further sensitizing applications (patch test negative or questionable).

We then sensitized the 18 guinea pigs to DNCB (see technique p. 4).

All the guinea pigs became sensitive to DNCB: the 9 animals originally strongly sensitive to PNDMA, gave strongly positive (macroscopically and histologically) patch tests on the flank with DNCB; the 9 which were less, or not at all, sensitive to PNDMA gave much weaker responses to DNCB under macroscopic examination, but under histological examination the discrepancy between the 2 groups was not marked.

On the contrary, histological examination of the nipple shows many more lesions in the first group (numerous spongiosis and significant infiltrate) than in the second.

Hence, it seems that animals poorly sensitized to PNDMA would be less capable of being sensitized to DNCB than those who initially are well sensitized with the first eczematogen (PNDMA).

*Macroscopic results on the flank and histological results on the nipple would confirm the clinical observations in man that certain individuals possess a greater capacity to being sensitized than others (eine Ekzembereitschaft, BLOCH, 1923).*

### 3. Citraconic Anhydride (CA) Followed by Dinitrochlorobenzene (DNCB)

We successively caused sensitization with CA and then with DNCB (technique p. 32 and 4). Patch tests with CA on one flank and DNCB on the other are both positive under macroscopic examination.

We then wondered if there would be a histological difference between the 2 tests, since, as we have seen, the infiltrate with the CA test in guinea pigs sensitized only to CA, is largely composed of eosinophils; while the infiltrate of the test with DNCB in guinea pigs sensitized only to DNCB is composed largely of small round cells (lymphocytes).

In the histological examination of the test with DNCB in guinea pigs sensitized to CA and DNCB, we found more eosinophils in the infiltrate than we did in that of the guinea pigs sensitized only to DNCB. Moreover, the number of eosinophils is greater than on the untreated skin of guinea pigs sensitized only to CA, but lower than that in a patch test to CA in this same guinea pig.

We should continuously keep in mind that guinea pigs sensitized to CA and subsequently to DNCB have a blood eosinophilia which is due to the first sensitization with CA (p. 45). We believe that some of these blood eosinophils are then fixed at the site of the patch test with DNCB.

# III. Hematological Observations (Dinitrochlorobenzene, Citraconic Anhydride and Propionic Anhydride)

We have previously pointed out (p. 33 and 36) that with certain eczematogenic substances the infiltrate contains eosinophils; we then wondered if there were modifications in the composition of the blood during the course of and following sensitization.

## A. Eosinophils

In normal, male albino guinea pigs we found in 12% of the animals (16 out of 128 guinea pigs) a blood eosinophilia of more than 3%.

In guinea pigs sensitized (for as long as 15 days or more after the sensitization period):

a) with DNCB, we never noted an increase in the eosinophilia and never or only rarely observed eosinophils in the infiltrate of the patch tests.

b) with CA, 14 out of 18 guinea pigs showed a blood eosinophilia greater than $3\%$ as well as numerous eosinophils in the infiltrate. The proportion of eosinophils in the blood is directly related to the proportion of eosinophils in the infiltrate of the patch test. The correlation coefficient is significant ($p < 0.005$) (N. HUNZIKER, 1965).

If the blood smear is repeated in the course of the months following sensitization to CA, the number of blood eosinophils remains higher than that of control animals of the same weight. Furthermore, the patch tests also remain positive, always revealing numerous eosinophils.

After 1—4 hours, a patch test with CA in guinea pigs sensitized to this substance, induces a momentary drop in blood eosinophils followed, after 12— 14 hours, by an increase which often results in a greater number of eosinophils than there were before the test. This decrease in the number of eosinophils is significant ($p < 0.001$ and the increase which follows is also significant ($p < 0.005$).

In guinea pigs sensitized to CA, we even noticed an increase in eosinophils in the regions of the skin which were not treated, something we never saw in the skin of a non-sensitized guinea pig or one sensitized to DNCB. However, this generalized cutaneous eosinophilia is considerably inferior to that localized at the site of the patch test.

c) with PA, about $50\%$ of the guinea pigs had a blood smear showing more than $3\%$ eosinophils. In the infiltrate of the patch tests, there was a great variation in the number of eosinophils found.

However, the infiltrate was not composed mostly of eosinophils.

# B. Basophils

In normal, male albino guinea pigs (107 animals) we find an average of $0.6\%$ of basophils in the blood. This figure corresponds to that found in the literature (see for exemple MICHELS, 1938).

In guinea pigs sensitized to chemically simple substances (DNCB, CA, PA), we never observed a blood basophilia in the very numerous blood smears done, two weeks after sensitization, and throughout the following months. Likewise, we noted nothing following a patch test.

On the contrary, at the end of the sensitization treatment:

a) with DNCB (28 guinea pigs) we observed a distinct increase in the number of basophils, the average being $3.9\%$ (Fig. 23).

b) with CA (9 guinea pigs) we find an increase in blood basophils, the average being $3\%$.

c) with PA (34 guinea pigs) we also found a clear increase in blood basophils, the average being $4.05\%$. Some guinea pigs even showed a basophilia of more than $7\%$, one showed as much as $25\%$.

Fifteen days after the period of sensitization (DNCB, CA, PA), the basophilia had disappeared.

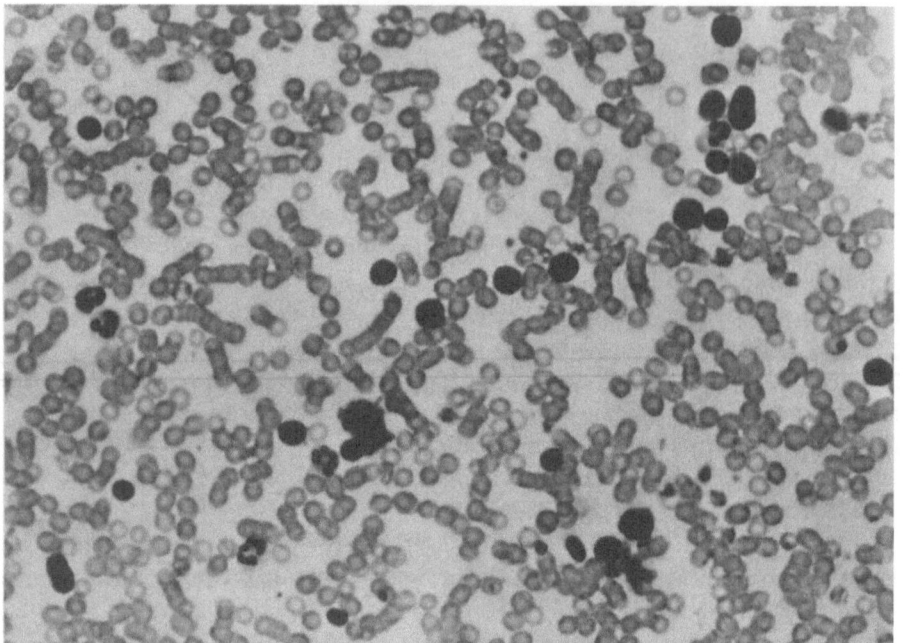

Fig. 23. Blood smear of a guinea pig at the end of a sensitization treatment with DNCB:
many basophils

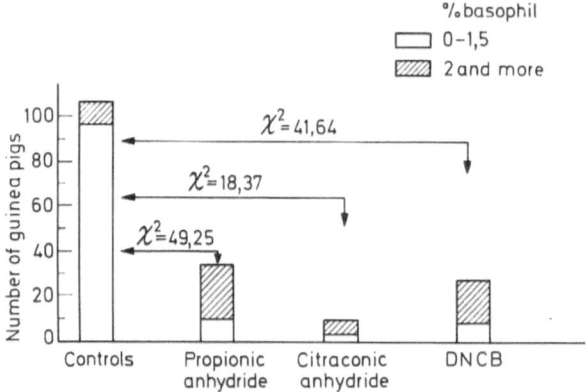

Fig. 24. Basophilia in control (non-sensitized) guinea pigs and in treated guinea pigs (near
the end of sensitization)


*Thus, this basophilia, which is observed towards the end of the sensitization period is a
transient phenomenon* (N. HUNZIKER and coll., 1965).

If the animals from each group, controls and sensitized guinea pigs (DNCB, CA,
PA), are divided into 2 classes, those having a percentage of blood basophils between
0 and 1.5% and those having a level of 2% or greater, the difference between the
control group and the 3 groups of sensitized guinea pigs is highly significant
($\chi^2$ method) (Fig. 24).

In 12 guinea pigs, we studied the evolution of this basophilia throughout the course of the treatment with PA.

There was a significant increase on the 14th day which lasted until the 16th day. Between the 16th and 18th day the number of basophils dropped to its normal value. This increase in basophils is not regularly progressive (Fig. 25).

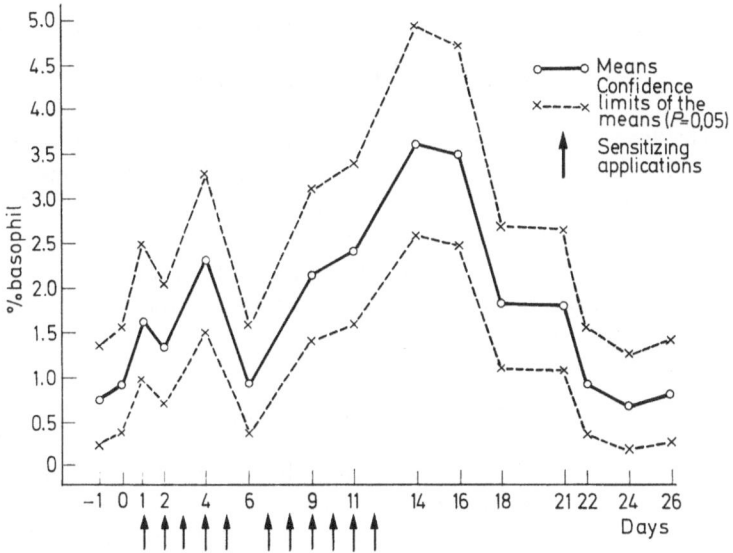

Fig. 25. Evolution of basophilia during sensitization with propionic anhydride (12 guinea pigs)

In another group of 11 guinea pigs, which was sensitized to PA but which is not included in Fig. 25, the increase in basophils found in the blood is already significant on the 10th day (an average of 3.6%). This increase was no longer evident 3 days later, that is on the 13th day or one day after the end of the sensitization period (average 1.9%).

As a control, some unsensitized guinea pigs were irritated on the neck 11 times in 12 days with a scalpel, and they received 4 subcutaneous injections of 0.05 cc of olive oil. At the end of this treatment, the necks are extremely irritated and yield the same clinical results as those following a sensitization treatment:

These guinea pigs showed no blood basophilia.

In 1960 WINQUIST induced a basophilia in guinea pigs between the 10th and 14th day by the injection of horse plasma. We call attention to the fact that this author's graph shows an almost significant increase in basophils on the 2nd day, which is followed by a drop in the graph, which is comparable to that recorded in our experiments.

# C. Lymphocytes

In some guinea pigs we also examined the percentage of lymphocytes throughout the course of the sensitization treatment (before, during and after), but we were not able to demonstrate significant variations. This, agrees with the observations of SEEBERG (1953), who similarly did not observe significant variations in lymphocytes during the sensitization of guinea pigs to DNCB.

Summary: *the increase in eosinophils is a lasting phenomenon, differing in intensity with CA and PA, but non-existent with DNCB; while the increase in basophils is a transient phenomenon common to the 3 eczematogens with which we experimented. We do not observe significant variations in lymphocytes throughout the course of the sensitization treatment.*

# IV. Comparison and Discussion of the Results of the Various Sensitizations

It appears, from the studies of other authors and from our own experiments, that it is possible to sensitize guinea pigs with various chemically simple substances.

However, in addition to group or individual variations and to possible seasonal variation, there are variations in the manner of sensitization and in the specific reaction of sensitization inherent in the substance itself. In fact, though most of the substances we examined are simple eczematogens, one of them, CA, is successively urticariogen and eczematogen.

## A. Macroscopic Test

Under macroscopic examination it is often difficult to differentiate between the different tests (according to the substance used).

However, *macroscopic distinction between the different eczematogens is occasionally possible in following the evolution of the test with time (time of appearance, state and disappearance of the test).*

In guinea pigs sensitized to CA, an *urticarial* reaction often appears 30 minutes after application of the substance. After 1 to 2 hours, this urticarial reaction is then followed by papules, which are generalized more or less over the entire body. The urticarial reaction due to CA progresses into a delayed, eczematous reaction, exactly like that seen in a test with DNCB in guinea pigs which have been sensitized to this substance.

With PA, we did not observe a definite urticarial reaction*. With DNCB, PNDMA, and potassium bichromate we never observed an urticarial reaction.

Thus, it is possible to distinguish in the macroscopic examination of the test, the guinea pigs sensitized to CA, from those sensitized to another substance by the urticarial reaction.

Moreover, about 14 hours after the test we are no longer capable of distinguishing between the patch tests performed with the various substances we employed. However, we have observed differences in the *disappearance time* of the reaction. The erythema due to CA diminishes rapidly after 14 hours; that due to DNCB and PNDMA diminishes quite quickly after about 24 hours; that due to potassium bichromate diminishes more slowly, but in this case the test is performed by intradermal injection and not by epicutaneous application. On the other hand, the intensity of the test with PA continues to increase between 14 and 24 hours until the epidermis actually gives the impression of potential erosion, but this phenomenon subsides and the reaction slowly diminishes, more slowly than with DNCB.

---

* However, we did not inject a stain to show this reaction.

We would like to add that there is a certain amount of subjectivity to the macroscopic appreciation of the test. This is one of the reasons why histological examination should be carried out as often as possible.

# B. Histological Examination

A histological examination provides us with more information as to what is happening during a patch test. In the epidermis there are spongiosis, swollen cells, acanthosis, etc. . . . In the dermis, there is an infiltrate and edema. The following differences are observed according to the allergen used:

a) In guinea pigs sensitized to DNCB we observe spongiosis of the flank as well as of the nipple. This spongiosis may or may not contain lymphocytes (inhabited or uninhabited spongiosis).

In guinea pigs sensitized with CA, PA and PNDMA, spongiosis are less frequently found. They are present almost always on the nipple, but rarely on the flank.

b) In some guinea pigs sensitized to PNDMA, we recorded acanthosis on the flank, but no spongiosis. Would acanthosis be a sign of sensitization in this case? Does the flank of a sensitized animal respond to the allergen either by spongiosis or by acanthosis?

c) In guinea pigs sensitized to PA, we observed epidermal lesions on the flank which were so intense that one could describe the intra-epidermal vesiculation as progressing to a necrosis: This reminds one of Lyell's disease (toxic epidermal necrolysis). They are not primary toxic lesions but lesions consequent upon sensitization. In these same guinea pigs, the nipple can show spongiosis.

We should like to insist again that, in guinea pigs which have received a sensitization treatment with paraphenylenediamine, all the patch tests were negative under macroscopic examination. However, in one case we observed well developed spongiosis on the nipple. MACHER and SENNLAUB (1963) have confirmed this observation with DNCB. This again demonstrates that microscopic examination affords more information than macroscopic examination.

d) The infiltrate is important in differentiating between the different eczematogens. In the tests with DNCB and often with PNDMA (in guinea pigs sensitized to these substances), the infiltrate is generally moderate, composed especially of small, round cells (lymphocytes) in or around the vessels. After 5 hours they invade the spongiosis (inhabited spongiosis), although there are still uninhabited spongiosis which can also be seen. In one group of guinea pigs sensitized to PNDMA, we found an infiltrate very rich in eosinophils. These eosinophils were not present in the other groups sensitized in exactly the same manner. The infiltrate observed in the reaction with CA shows few lymphocytes, but already after 30 minutes, numerous eosinophils can be seen which increase in number in the hours that follow and invade the epidermis. These eosinophils compose the greater part of the infiltrate.

In the sensitization with PA, the infiltrate is very extensive. It is composed of lymphocytes and eosinophils, but the latter do not constitute the greater part of the infiltrate as they do with CA.

In the sensitization with potassium bichromate, we obtained clear positive reactions only with the intracutaneous tests. The infiltrate is extremely dense, but does not penetrate the epidermis and it contains no eosinophils. Furthermore, we did not observe clear epidermal lesions. In this case, apparently only the dermis participates in the sensitization reaction (dermal sensitivity, see p. 70).

In conclusion, although it is often difficult to distinguish macroscopically between the reactions due to the different eczematogens, this distinction becomes easier under microscopic examination when one considers the differences in the epidermal lesions (spongiosis, acanthosis, necrosis), the quality of the infiltrate, and the magnitude of this infiltration.

The histological examination is relatively objective. The slides are retained and permit a repetition of the examination, a comparison, and a discussion of the results. This is why we attach such a great significance to the histological examination.

## C. Hematological Variations

In blood smears we can observe differences similar to those seen in the infiltrate.

In guinea pigs exhibiting an infiltrate which is rich in *eosinophils*, after a patch test with CA and also often with PA, we can also observe, a blood eosinophilia. This is not seen in guinea pigs sensitized to DNCB.

*Thus, there is a correlation between the quantity of tissue eosinophils after the test and that of blood eosinophils.* This relationship is significant.

The influence of the blood cell composition on the cutaneous infiltrate is established. For example, if a guinea pig has undergone a double sensitization (CA followed by DNCB), there is a blood eosinophilia due to the first sensitization (CA). Then after a test with DNCB, the infiltrate contains some eosinophils, an observation never made following a test with DNCB in a guinea pig sensitized only to the latter substance.

We verified another hematological variation which seems common to diverse sensitizations (DNCB, CA, PA): the transient increase in blood *basophils* between the 10th and 16th days.

This transient basophilia seems to be connected with a more general phenomenon.

Already in 1902 LEVADITI described a basophilia in the guinea pig and rabbit during the course of sensitization with foreign proteins. SCHLECHT (1909/10) and WINQVIST (1960) also reported an increase in basophils following an injection of serum or heterogeneous plasma into guinea pigs.

In the course of the sensitization of man to *Novobiocine*, a basophilia, interpreted as an immunological response, has been observed by SHELLEY (1963).

*Thus, the increase in basophils that we observed links sensitization with chemically simple substances to a larger phenomenon, that is, sensitization in general.*

# V. Remarks on Passive Transfer

Passive transfer of the eczematous sensitivity is very important in understanding the mechanism of the delayed eczematous reaction, all the more so since until now, it has not been possible to prove the existence of antibodies in the serum, in either man or animals. This impossibility has led certain authors (KARUSH and EISEN,

1962) to believe that the concentration of antibodies does not reach a level which is detectable by current immunological techniques. Other authors believe that the antibodies are bound to the cells and are transferred by the cells and not through the serum (LANDSTEINER and CHASE, 1942).

a) The first transfer tests were attempted in 1925 by URBACH. He injected healthy subjects with the contents of eczematous blisters and subsequently tested the sites of injection.

With this method of KÖNIGSTEIN-URBACH (1921), several authors obtained positive results in man (FELLNER, 1919, 1933) and animal (PERUTZ (1928), FUHS and RIEHL (1928), BRANDT and KONRAD (1930), BIZZOZERO and FERRARI (1931), BALLESTERO and MOM (1945), etc.). With the same method, other investigators obtained negative results, casting doubt on this technique (W. JADASSOHN, 1932; SULZBERGER, 1940; LEIDER and BAER, 1948 etc.).

b) In 1942, LANDSTEINER and CHASE established a passive cellular transfer technique: they inject cells from strongly sensitized animal donors. In this way, they were able to transmit an eczematous sensitivity to DNCB and picryl chloride to animals which had not been previously sensitized. CHASE (1953) demonstrates a specific tuberculin hypersensitivity in normal guinea pigs by the injection of leukocytes isolated from peritoneal exsudates and leucocytes from the blood, spleen or lymph nodes.

With guinea pigs, this technique has been successfully employed by many authors in transmitting sensitivity to tuberculin (CHASE, 1945; METAXAS and METAXAS-BÜHLER, 1948 and 1954), or contact eczematous sensitivity to chemically simple substances (HAXTHAUSEN, 1947, 1951 and 1952; CREPEA and COOKE, 1948; NILZEN, 1952; SERRI, 1954; SKOG, 1955; GRIMMER and SPIER, 1961; GROTH 1963; KIND and coll., 1965; GUTHRISE and coll., 1966; SCHRÖPL and coll., 1966 etc.).

If cellular passive transfer is easily reproducible in guinea pigs, it is much more unreliable in man. Several authors have obtained negative results (HAXTHAUSEN, 1947 and 1952; BAER and SULZBERGER, 1952; BAER and coll. 1952; MENEGHINI and LEVI, 1959; SERRI, 1959; HARBER and BAER, 1961 etc.). However, the studies of W. L. EPSTEIN and KLIGMAN (1957) concerning the transfer of sensitization with chemically simple substances and the studies of LAWRENCE (1959) concerning the transfer of sensitization with tuberculin and hemolytic streptococci have been positive.

Cellular passive transfer gives more positive results when the number of injected lymphocytes is great (SKOG, 1955). It is successful even if the lymphocytes have been lysed by sonic vibration (JETER, TREMAINE and SEEBOHM, 1954).

In our clinic, DE WECK and BRUN (1956) did not succeed, at first, in transmitting sensitivity to DNCB, but on the contrary, guinea pigs which had intraperitoneally received spleen cells from animals strongly sensitized to picryl chloride, macroscopically showed strongly positive reactions after application of this eczematogen. But these animals were sensitized with Freund's adjuvant (according to CHASE) while the sensitization of animals to DNCB was by epicutaneous application. Histologically, however, epidermal lesions on the flank and nipple were not observed in the recipients. This is the opposite to what was observed in the donor animals. Hence, sensitization appears to have been transmitted by this procedure; but we cannot say if it is really a complete sensitization, since the histological examination disclosed no epidermal lesions. However, GRIMMER and SPIER (1961) were able to

4*

transmit sensitivity to DNCB which was confirmed by histological examination. They observed epidermal and dermal lesions in the recipient guinea pigs as well as in the donors. Nevertheless, their method of sensitization (epicutaneous application of the eczematogen) was different from that of DE WECK and BRUN* (1956) in our clinic.

However, according to SPIER (1965) this difference in sensitization cannot explain everything.

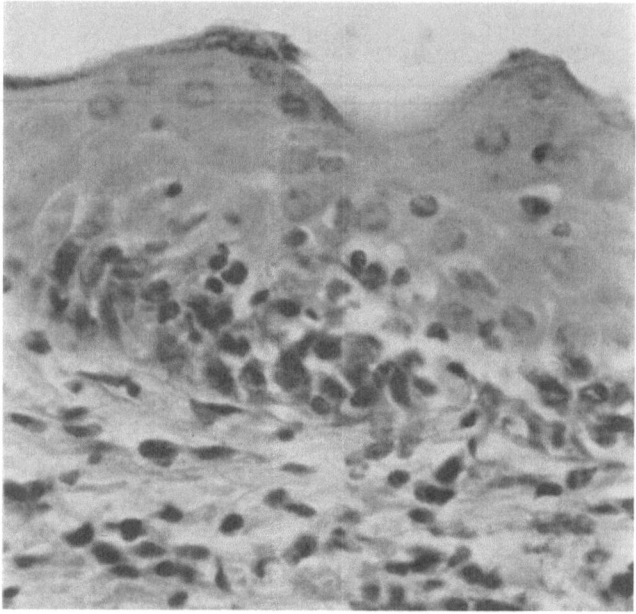

Fig. 26. Guinea pig non-sensitized; passive transfer by serum from a guinea pig sensitized to CA. Flank 24 hours after a single application of CA 25% in dioxane: little spongiosis

In sensitizing guinea pigs by epicutaneous applications of DNCB, GROTH (1963) was able to transfer sensitivity to other animals. In the recipients, histological examination reveals an insignificant change. NAJARIAN and FELDMAN (1963) also fail to show in the recipients an image which is typical of contact eczema (donor animals sensitized to dinitrofluorobenzene).

*By transmitting lymphocytes from a sensitized to an unsensitized guinea pig, it seems possible to transmit an eczematous sensitivity in certain cases and for a certain time.*

Thus, attention has been attracted to the role of lymphocytes by CHASE's method. There is no clear understanding how the transferred lymphocytes provoke the cutaneous reaction in the recipient. This question is discussed by HAGERMAN (1954), NAJARIAN and FELDMAN (1961, 1962 and 1963), TURK (1962), TURK and ASHERSON (1962), HAMILTON and CHASE (1962), McCLUSKEY and coll. (1963), MENEGHINI and coll. (1963 and 1965) as well as by other authors.

---

\* Sensitization of animal donors with Freund's adjuvant according to CHASE.

This role of lymphocytes has been the subject of numerous studies in relation to the delayed type of allergic reaction, but this is not the place for analysis of this vast, but very interesting problem.

*Passive transfer of the sensitization to CA.*

*From guinea pigs sensitized to CA, we were able to transmit the sensitization to normal animals, by a modification of the Praunsnitz-Küstner method** (N. HUNZIKER, 1964). It has been known for a long time that an immediate type of sensitization can be transmitted with serum. Indeed, in guinea pigs sensitized to CA, we have seen that the delayed reaction is preceded by an urticarial reaction. Thus, it is not surprising that we were able to transmit a reaction of the immediate type by serum, which is visible after 30 minutes (as CHASE 1947 described with this same substance).

However, we have seen that the transferred immediate urticarial reaction is progressively transformed, in the recipient as well as in the donors, into a delayed reaction which reaches its maximum about 8 hours after the test, but which is still visible after 24 hours.

From the histological point of view, the urticarial papule, due to the passive transfer, shows an infiltrate in the dermis which is very largely composed of eosinophils: after 24 hours the delayed reaction shows some rough outlines of spongiosis, an acanthosis, and an infiltrate largely composed of eosinophils (Fig. 26).

*Consequently, it would seem that an intradermal injection of serum from sensitized guinea pigs into normal guinea pigs causes the transmission of an urticarial reaction which changes into a delayed type of reaction, even if this reaction is not a strong one.*

# VI. Influence of Various Factors on Sensitization

## A. Influence of Corticosteroids

At the present time, there is no doubt that a patient suffering from a serious eczema is spectacularly improved by corticosteroids.

However, SULZBERGER and BAER (1951) have called attention to the fact that in man, patch tests are barely affected by corticosteroids, even with treatments of long duration.

Thus, there is a temptation to prove this experimentally. Numerous investigators have tried to do so, but for the most part, the results have been disappointing.

Cortisone has no effect on sensitization to DNCB and the initiation of the patch test (BALDRIGE and KLIGMAN, 1951; FREY and STUDER, 1951). „Es wurde eine gewisse Hemmung der Sensibilisierung festgestellt, doch ließ sich die Sensibilisierung nicht verhindern" (MIESCHER and SONCK, 1952).

Macroscopically, HITCH (1953) reported no difference among guinea pigs sensitized with DNCB, some of which were treated by corticosteroids. Under histological examination the reaction of the treated guinea pigs appears weaker; however, this difference is not significant.

NILZEN (1952) has reported a decrease in guinea pig sensitivity to DNCB by systemic treatment with cortisone.

---

* From guinea pigs sensitized to PA, we were not able till now to transmit the sensitization to normal animals, by the same method.

In guinea pigs sensitized to DNCB, we have not been able to show an influence of cortisone in the post-eczematous increase of mitotic activity (BUJARD, W. JADASSOHN and MUSSO, 1954).

1. We have attempted to establish the most favorable conditions for showing the action of corticosteroids. For this reason, we chose to sensitize the guinea pigs by applications of 0.1% DNCB on the left nipple for 20 days. Very high doses of corticosteroids were administered orally before, during and after this sensitization (N. HUNZIKER, 1961).

For prednisone, the dosage was 2.5 mg/day per guinea pig or 8.4 mg/kg; for triamcinolone the dosage was 2 mg/day per guinea pig or 6.3 mg/kg. For a man of 60 kg, the corresponding dose would be approximately 500 mg/day.

We examined 3 groups of guinea pigs:

1) 11 *animals*.

Preliminary treatment with prednisone for 8 days followed by a sensitization treatment for 20 days while continuing prednisone.

2) 11 *animals*.

The same treatment as for group 1 except that the prednisone is replaced by triamcinolone.

3) 13 *animals*.

No treatment with corticosteroids, sensitization treatment for 20 days.

Results: on the flank, macroscopic examination of the tests revealed no differences among the 3 groups.

On the 21st day, DNCB was applied to the 2 nipples of the guinea pigs of the 3 groups.

Histological examination of the nipples of the 3 groups of guinea pigs showed a picture typical of eczema in all the animals.

Thus, if guinea pigs are treated with prednisone or triamcinolone, before, during and after sensitization to DNCB, eczema can be induced, as it can in the controls (macroscopic and microscopic examination).

In conclusion, *the negative results we recorded concerning the influence of corticosteroids on eczema as well as other affections\*, do not allow us to say that cortisone is an anti-inflammatory hormone* without adding as did MUSSO (1956), *that it is anti-inflammatory in certain cases and under certain conditions.* We also used corticosteroids in guinea pigs sensitized to CA which show a distinct blood eosinophilia (see p. 44).

We examined 2 *groups of guinea pigs:*

1) Guinea pigs sensitized to CA and receiving intra-peritoneal injections of hydrocortisone disuccinate (25 mg/animal from 300—400 gr).

2) Guinea pigs sensitized to CA and receiving intra-peritoneal injections of solvent only.

In the guinea pigs of the first group, the injection of hydrocortisone produces a drop in the number of eosinophils in the blood from 0—0.5% within 6—8 hours.

---

\* Mycosis of the guinea pig (W. JADASSOHN, MACH, NARDIN, 1951), herpetic and vaccinal infection of the rabbit's cornea, tuberculin ophthalmoreaction of bovids, urticarial reaction in the Prausnitz-Küstner test, cutaneous erythema by UV rays (FRANCESCHETTI and coll., 1951), dermatitis to croton oil, Schick test in guinea pigs (MUSSO, 1956).

Three to seven days later, the percentage of eosinophils has returned to its initial value.

In the guinea pigs having received only the solvent, the percentage of eosinophils did not change.

On some of the guinea pigs sensitized to CA, we performed a test with CA when the percentage of eosinophils was between 0 and 0.5%.

Under macroscopic examination, these tests were positive; there was no difference between the guinea pigs of the group which were sensitized and which did not receive hydrocortisone.

We performed a biopsy in these patch tests after 14 hours, when the percentage of eosinophils was still between 0 and 0.5%. In the infiltrate of the tests, we observed numerous eosinophils which were intravascular, perivascular and disseminated in the dermis.

*Thus, we observed neither a macroscopic nor a histological difference between the tests of the guinea pigs having received hydrocortisone and those having received only solvent* (in other words, between the guinea pigs having and those not having an eosinophilia).

Moreover, in the animals which had received hydrocortisone, we noted numerous eosinophils in the dermis of some regions of the skin which were untested. It even appeared that these eosinophils were more numerous than in the skin of the guinea pigs which were sensitized to CA, but which had received no hydrocortisone.

However, in 2 patients having a very low percentage of blood eosinophils due to cortisone, HOFER and GOLAY in our department (1958) had shown that there was a decided reduction of eosinophils in the urticarial papule induced by morphine. In another case, where the blood eosinophilia was not affected by cortisone, there was no change in the number of eosinophils in the infiltrate of the urticarial papula. From these observations made on man, it would thus seem that there is a direct relationship between the percentage of eosinophils in the blood of the patients and the number of eosinophils in the urticarial papule.

This relationship is not confirmed by a patch test in guinea pigs sensitized to CA and treated by hydrocortisone. The case of HOFER and GOLAY was a matter of an *immediate reaction* in man, while we examined a *delayed reaction* in the guinea pig.

# B. Influence of X-Rays

According to a great many investigators, superficial radiotherapy can be very beneficial for eczema in man, especially in severe and chronic cases. However, SULZBERGER and ROSTENBERG (1939) have insisted that the true value of radiotherapy has not been established by critical methods. In 1954 KEMP and KLIGMAN observed no effect of X-Rays on contact eczema in man.

In the dermatological clinic of Geneva, ADATTO (1962), MAGGIORA and LOZERON (1962), LOZERON (1964), MAGGIORA, BUJARD and W. JADASSOHN (1967) attempted to observe the action of X-rays* on the eczema in guinea pigs due to DNCB (since such experimentation is not practically possible in man):

---

* The X-ray were administered with the Dermopan apparatus (position IV) 50 KV, 25 mA, filter 1 mm Al.

### 1. Can X-Rays Prevent Sensitization?

Guinea pigs are sensitized by repeated applications of 0.1% DNCB on the nipple (p. 5). The nipple was irradiated before sensitization either by a single dose of (1600 R), or by a fractionated dose (20 × 100 R: 2000 R).

The irradiations did not inhibit the sensitization of the guinea pigs.

### 2. Do X-Rays Prevent a Positive Test in a Sensitized Guinea Pig?

Even 800 R does not prevent a positive test.

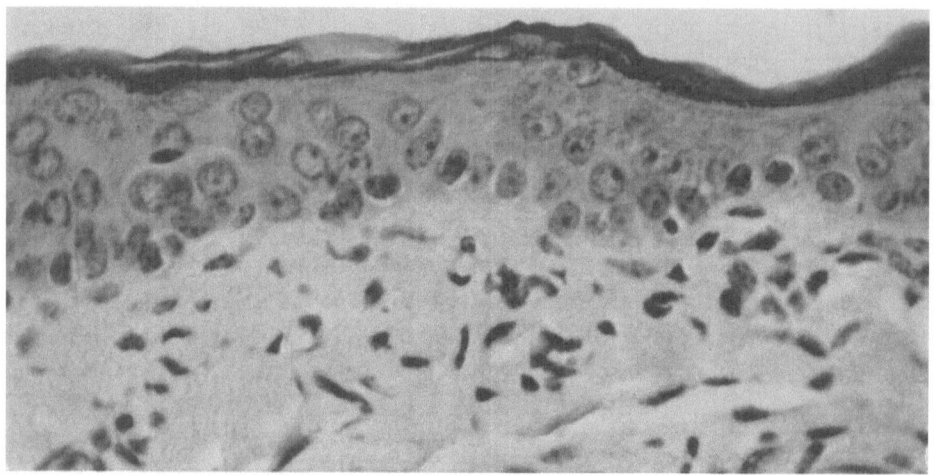

Fig. 27. Guinea pig sensitized to DNCB. Nipple irradiated (800 R) and excised 81 hours afterwards: no epidermal alterations

### 3. What is the Effect of X-Rays on an Eczematous Lesion?

a) 800 R applied to a flank without a patch test produces no lesions (Fig. 27). 800 R administered to a patch test on the flank of a guinea pig sensitized to DNCB provokes visible epidermal alterations (Fig. 29) which are different from those of a simple eczema as shown in Fig. 28. In the irradiated and patch-tested skin we observe: greatly swollen epidermal cells, cell walls often entirely disappeared, nuclei of varying sizes and greatly decreased in number; disordered basal layer; decrease in the number of mitosis (shown by colchicine, see p. 11) (Fig. 29). These changes resemble those visible later on the epithelium of the nipple which has received a dosage of X-rays 3 times as great (2400 R) and which appear much later after irradiation (W. JADASSOHN and coll., 1950; W. JADASSOHN and BUJARD, 1952).

b) What is the effect of X-rays administered immediately before the initiation of the test?

800 R is administered to the nipple of a sensitized guinea pig before the initiation of the test. Histological examination shows the same epidermal alterations as described above.

These particular epidermal alterations, observed either on the flank or on the nipple (whether or not irradiation had been carried out before or after the initiation of the test) are due to the combination of X-rays and eczema, since they are observed neither following the same dosage of X-rays alone nor following a patch test alone.

*The alterations which can be attributed to a combination of 2 factors (X-rays and eczema) can be termed "Kombinationsschaden".*

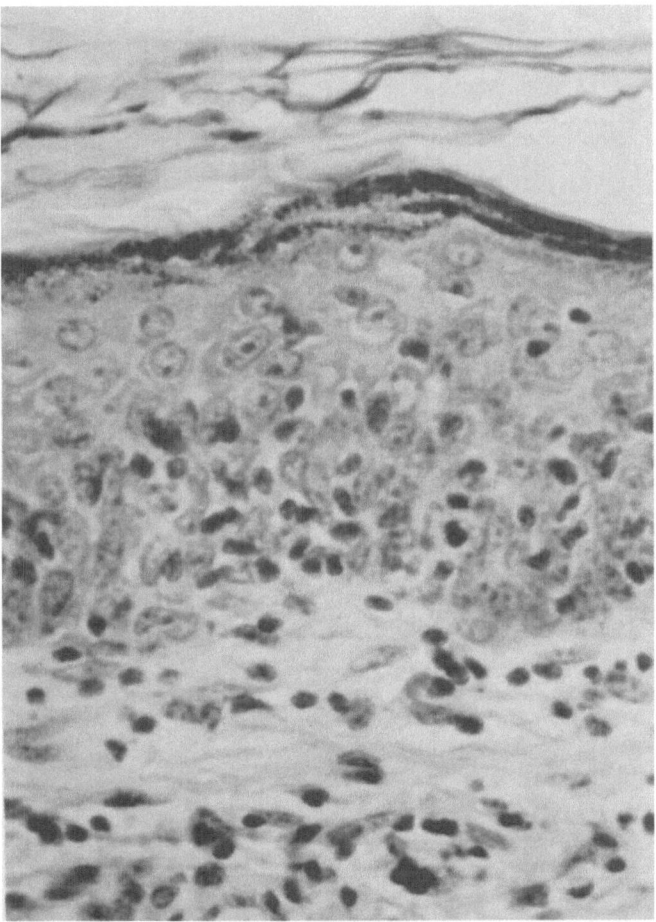

Fig. 28. Guinea pig sensitized to DNCB. Nipple 81 hours after a single application of 0.1% DNCB in acetone: spongiosis

## 4. What Occurs if a Patch-Test is Applied on the Irradiated Flank (1600 R) of a Guinea Pig?

It is known that this dosage of X-rays, when applied to the skin of the flank of the guinea pig, produces a definite epilation (MAGGIORA, 1961; MAGGIORA, BRUN, 1962). After several months, the skin of these guinea pigs is practically normal in appearance, disregarding the alopecia and the acanthosis which are still distinct after 61 days following the irradiation.

On the guinea pigs sensitized to DNCB and then irradiated, a test is performed on the irradiated flank (62 days after the X-rays), as well as on the flank which was not irradiated.

Macroscopically, the irradiated flank gives practically no responses, whereas the non-irradiated flank gives a positive reaction, as usual, to DNCB. Histological examination confirms the macroscopic observation (MAGGIORA, LOZERON, 1962).

*It seems therefore, that irradiation of the skin with a certain dosage protects the skin against the effect of an eczematogen.*

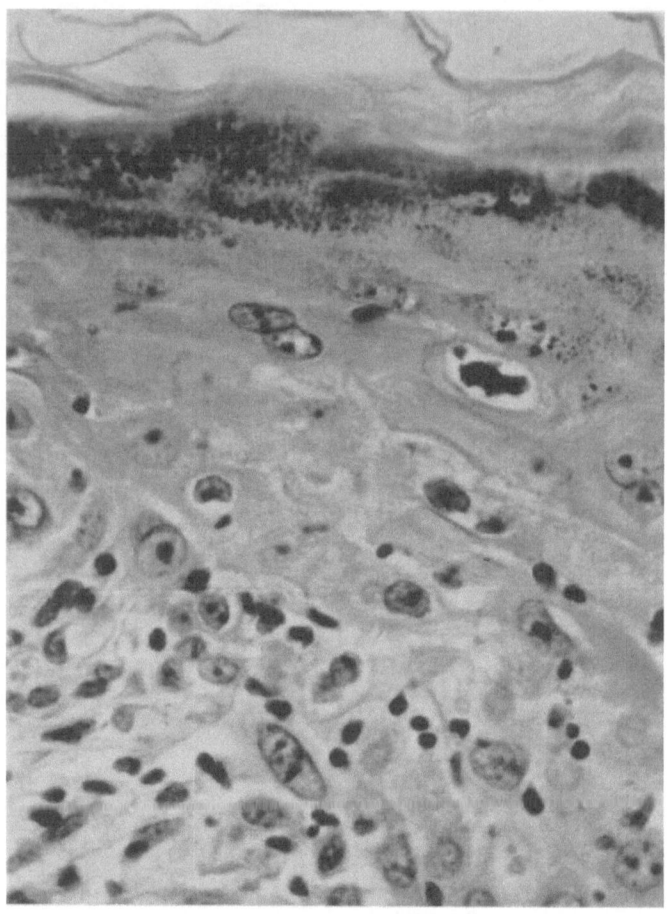

Fig. 29. Guinea pig sensitized to DNCB. Nipple irradiated (800 R), followed by a single application of 0.1% DNCB in acetone; nipple was excised 81 hours afterward: epidermal cells are extremely swollen, there are atypical cells

Before being able to affirm that irradiation directly prevents the appearance of eczema, one must establish the significance of the absence of hair follicles and of acanthosis.

a) To eliminate the role of hair follicles one nipple was irradiated with 1600 R. Nipples possess practically no hairs. The histological results show an eczematous reaction which is less extensive on the irradiated side than on the non irradiated side.

Hence it is not possible to explain the weak eczematous reaction on the irradiated flank by deficiency of hairs.

b) The irradiated flank shows an acanthosis. Therefore it is necessary to compare this flank when tested with the other flank also showing an acanthosis, not produced

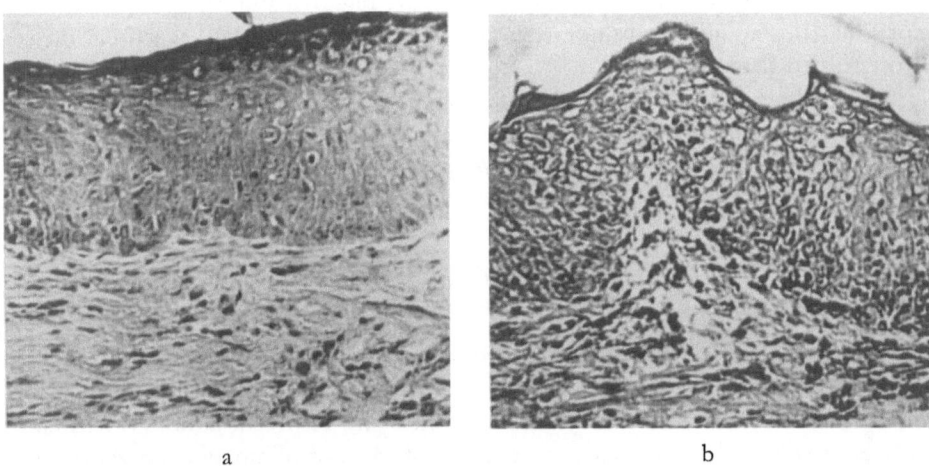

a                                                        b

Fig. 30. a Guinea pig sensitized to DNCB. Acanthosis of the flank (due to X-rays) 14 hours after a single application of 0.1% DNCB in acetone: a few epidermal alterations, no spongiosis. b Guinea pig sensitized to DNCB. Acanthosis of the flank (due to erucic acid) 14 hours after a single application of 0.1% DNCB in acetone: spongiosis

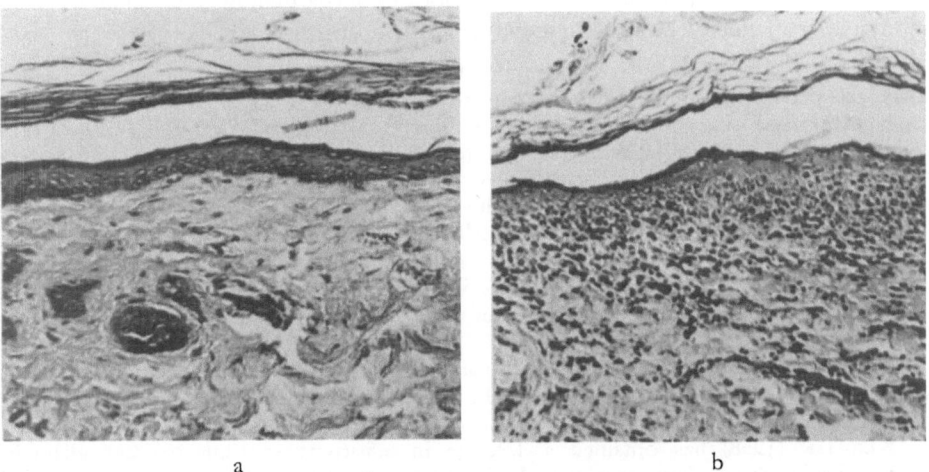

a                                                        b

Fig. 31. a Guinea pig sensitized to DNCB. Flank irradiated (1600 R) 90 days before the test: no acanthosis, no spongiosis. b Guinea pig sensitized to DNCB. Flank was not irradiated. After a test: spongiosis

by X-rays but by erucic acid. It appears from this experiment that the macroscopic and histological reactions are much weaker on the irradiated flank than on the flank whose acanthosis was caused by erucic acid (Fig. 30a and b).

After a sufficient lapse of time following an irradiation of 1600 R on the flank of normal guinea pigs, the acanthosis disappears and, except for the absence of hair follicles, the skin appears completely normal. If a patch test is performed after the guinea pigs have been sensitized, there is almost no macroscopic reaction, nor is there a histological reaction comparable to that observable on the non-irradiated flank (MAGGIORA, BUJARD, W. JADASSOHN, 1967, Fig. 31a and b). This proves once more that the acanthosis is not responsible for the diminished reactions on the X-ray treated flank.

To summarize, *in the sensitized guinea pigs, application of a certain dosage of X-rays (1600 R) inhibits the eczematous reaction triggered by DNCB. This inhibition does not appear to depend on the absence of hair follicles or on acanthosis, but on the direct effect of the X-rays.*

# VII. Desensitization

Already in 1923, J. JADASSOHN had discussed the problem of desensitization in man with respect to a case of eczema due to Odol. Since then, this problem has preoccupied all dermatologists and is far from being resolved, although its importance is growing with the increasing number of cases of occupational eczema.

Sensitization acquired during one's lifetime can, nevertheless, subsequently disappear. In some patients patch tests which were once positive have become negative. Masons who have suffered from eczema due to cement do not necessarily undergo a relapse if they take up their construction work again (HUNZIKER and MUSSO, 1960). Is there undeniably a diminution of eczematous hypersensitivity? Is this desensitization specific or not, is it localized (hardening?) or general, is it spontaneous or due to repeated contact with small doses of the allergen?

Moreover, according to SEQUEIRA and coll. (1947), HELLIER (1950), an acquired sensitivity can persist throughout life; but according to MORGAN (1953) this is not by any means necessary. Using eczematous people as controls, MIESCHER (1938) GOMEZ ORBANEJA and BARIENTOS (1938) found more negative tests in people who had avoided contact with the eczematogen than in the others. However, can these people really be said to have avoided all contact with eczematogens, such as turpentine for exemple, for all these years? MIESCHER (1938) does not exclude the possibility of desensitization by repeated contact.

When patients whose tests to nickel were positive were controlled, 43% of the tests came out negative: "The loss of sensitivity was generalized and could not be constantly related either to continued contact with possible hardening or to avoidance of further contact" (MORGAN, 1953).

For positive results in man, we will cite among others: WEDROFF and DOLGOFF (1935) who, with DNCB, showed that subjects, who are less strongly sensitized, are more rapidly desensitized.

KLIGMAN (1958) has obtained a decrease in sensitivity to Rhus toxicodendron in sensitized subjects who have received a homologue of the Rhus antigen.

TEES and MILNER (1960) have described a case in which there was a decrease in sensitization in eczema caused by nickel.

SOURREIL and coll. (1964), as well as FRUCHARD and FRUCHARD (1957), obtained a desensitization in eczema due to cement.

*In sensitized guinea pigs,* repeated tests diminish the degree of sensitization (for turpentine: BURCKHARDT, 1938; for DNCB: FREY, 1951).

GINGSBERG and coll. (1937) were able to desensitize guinea pigs which were sensitized to poison ivy but they dealt with only 2 animals.

# A. General Desensitization

After 100 days, macroscopic and histological examinations prove that guinea pigs sensitized on the neck are still sensitive to tests (0.1% DNCB in acetone) on the flank. There is no spontaneous disappearance nor clear decrease in the degree of sensitization (Fig. 32).

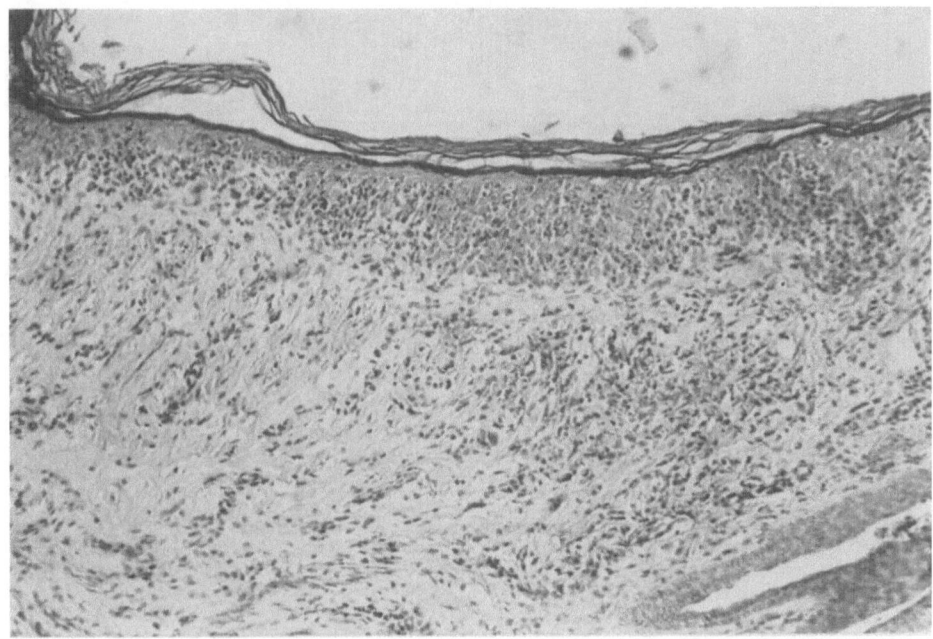

Fig. 32. Guinea pig sensitized to DNCB. No treatment for 100 days. Nipple after a single application of 0.1% DNCB in acetone: spongiosis

*After* 20 *daily applications* of DNCB (0.1% in acetone) on the nipple of a guinea pig sensitized to this substance by the usual method, there is no discernible desensitization.

*After* 40 *daily applications*, there is a slight desensitization, that is, the tests which are positive after 24 hours become negative after 48 hours, while they remain clearly positive in the control guinea pigs (guinea pigs sensitized to DNCB after 40 days of no treatment). Histological examination discloses no difference.

*After* 60 *daily applications*, desensitization in some guinea pigs is already unmistakable, although not complete (N. HUNZIKER, 1961).

*After* 100 *daily applications*, a group of guinea pigs gave a macroscopic reaction which was weak, questionable or negative. Histological examination does not show clear cut lesions in any of these animals (nipples and flanks) (W. JADASSOHN, BRUN and BUJARD, 1959) (Fig. 33) (Tabe 7).

In another group, 100 daily applications of DNCB to one nipple did not lead to a general desensitization, 150 applications were required. (The controls were still

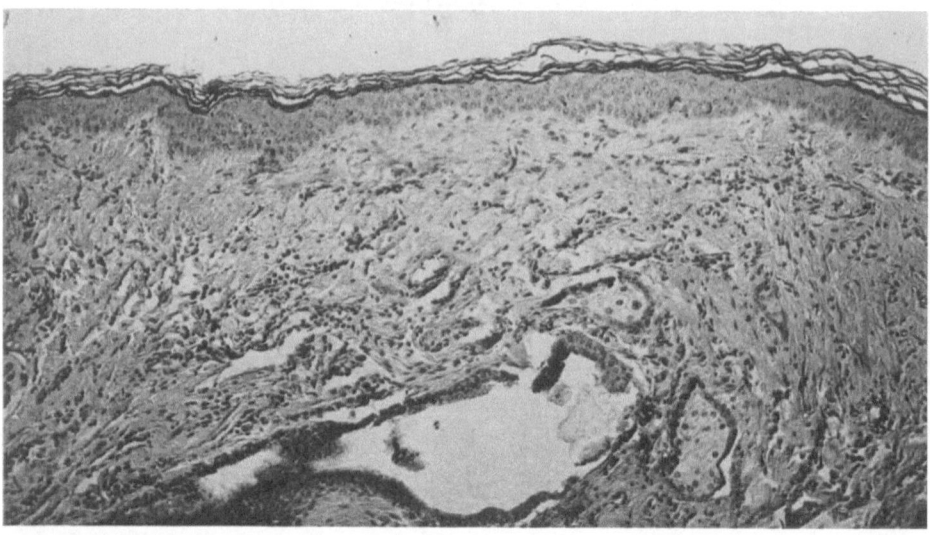

Fig. 33. Guinea pig sensitized to DNCB. Daily applications of 0.1% DNCB in acetone on the left nipple for 100 days. Right nipple after a single application of 0.1% DNCB in acetone: no lesions

Table 7. *Desensitization (DNCB) (one of our experiments)*

| Guinea pig No. | eczematous lesions of the right nipple 14 hours after one application of DNCB 0.1% in acetone | Treatment |
|---|---|---|
| 5934 | ± | Sensitization by 11 applications |
| 5935 | + | of 1% DNCB in acetone on the |
| 5936 | ± | neck. Then the left nipple is |
| 5937 | ± | treated 100 days with 0.1% |
| 5938 | + | DNCB in acetone |
| 5939 | ± | |
| 6361 | + | |
| 6362 | + | |
| 6363 | — | |
| 6364 | — | |
| 6365 | — | |
| 6366 | — | |
| 6367 | ± | |
| 6368 | — | |
| 6369 | +++ | Sensitization by 11 applications of |
| 6370 | ++ | 1% DNCB in acetone on the |
| 6371 | + | neck. No treatment for 100 days. |
| 6372 | +++ | |
| 6373 | ++ | |
| 6374 | ++ | |

positive.) This shows, once more, that there are differences between the various groups of guinea pigs (N. Hunziker, 1963).

*Our experiments show that it is possible to obtain a general desensitization by repeated applications to a very small surface (nipple)\*.*

Desensitization with DNCB has been partially confirmed by Rajka and Hard (1960). It has also been verified with DNCB, Taractan and PNDMA by Lowney (1964). It is specific, that is, guinea pigs sensitized to 2 substances and having received a desensitization treatment to only 1 substance, exhibit a weaker degree of sensitivity to this substance but not to the other. Employing our method, Schulz (1965) also demonstrated desensitization of the guinea pig sensitized to DNCB.

*Note:* We observed that guinea pigs can be sensitized by 20 daily applications of 0.1% DNCB in acetone to the nipple alone, but if these applications are continued from 60 to 100 days, the general sensitivity of the guinea pigs, instead of increasing, decreases. This observation corresponds to that previously reported in guinea pigs which are sensitized in the usual manner and then submitted to a series of applications.

# B. "Local Desensitization"

What occurs on the nipples of guinea pigs (previously sensitized by the usual method) which receive multiple applications of 0.1% DNCB in acetone?

a) *After 20 daily applications* (6 guinea pigs) there is a distinct acanthosis as well as eczematous lesions (spongiosis and infiltrate); while the other nipple, having received a single initiatory application, also manifests eczematous lesions which are more severe than those on the treated nipple.

b) *After 40 daily applications* (8 guinea pigs), there is a distinct acanthosis and few if any eczematous lesions (Fig. 34), while the other nipple, which received only one initiatory application, shows lesions typical of eczema (Fig. 35).

c) *After 60 daily applications* (14 guinea pigs), the nipple presents a clear acanthosis, but no spongiosis. The other nipple which received only one application, no longer exhibits spongiosis after a test with DNCB. Thus, a general desensitization has already taken place.

To summarize, *after* 40 *applications to one of the nipples, there is no general desensitization, but a "local desensitization" on the treated nipple* (N. Hunziker, 1961). It is not the acanthosis which has prevented the appearance of the lesions, since, as we have already stated (p. 19) acanthosis of the nipple due to estrogens does not prevent the appearance of spongiosis.

For the present, we can say that there is a "local desensitization", but we do not yet know whether or not it is specific.

---

\* Since then we have shown that it is also possible to desensitize guinea pigs sensitized with paranitrosodimethylaniline by the same method. The most striking demonstration of this desensitization is the almost total absence of eosinophils in the dermis, when the patch test is examined, on the contrary of the control group (Hunziker, 1968).

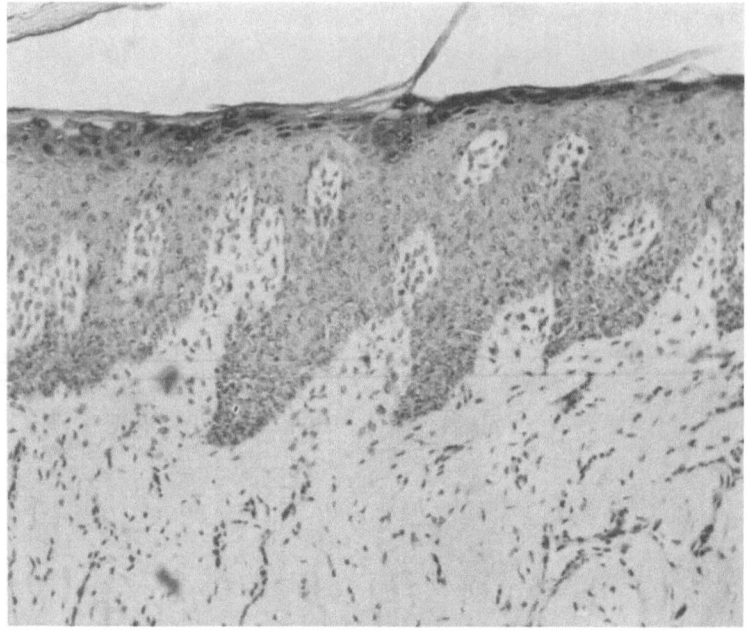

Fig. 34. Guinea pig sensitized to DNCB. Left nipple after 40 daily applications of 0.1%
DNCB in acetone: acanthosis, no spongiosis

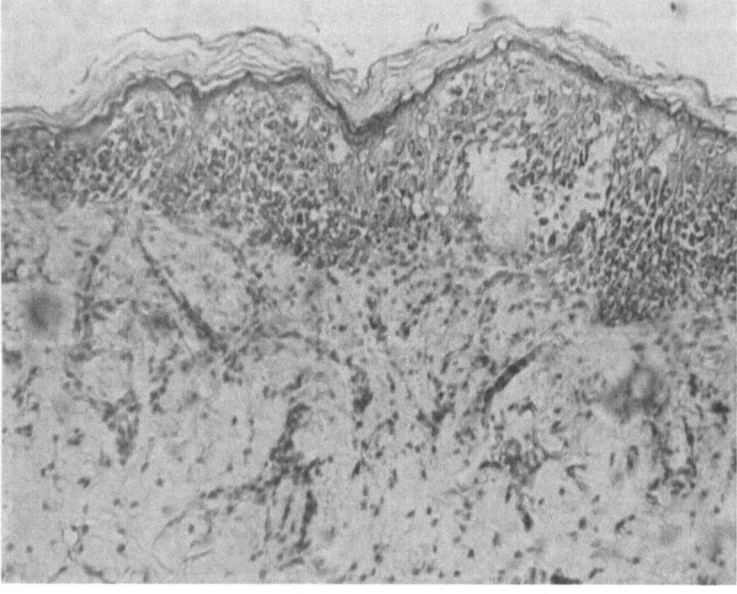

Fig. 35. Same guinea pig as above. Right nipple after a single application of 0.1% DNCB
in acetone: intraepidermal vesicles, spongiosis

# C. Duration of Desensitization and Resensitization

It would be interesting to know whether the acquired general desensitization is lasting. In any case, we can say that, in comparison to the control group, the desensitized guinea pigs are relatively less sensitive. This difference is macroscopically, as well as histologically verifiable.

LOWNEY (1964) confirmed that "hyposensitization" in guinea pigs lasts at least from 5 weeks to 2 months.

It is also interesting to establish whether desensitized guinea pigs can be resensitized with the same eczema-producing substance.

We took 3 groups of guinea pigs:
1. guinea pigs sensitized to DNCB and then desensitized to DNCB,
2. guinea pigs sensitized to DNCB,
3. unsensitized guinea pigs.

In each of the 3 groups we performed a sensitization treatment by our usual method:

After 4 sensitizing applications, we noted that the necks of the desensitized guinea pigs were much more irritated than those of the other guinea pigs.

At the end of the sensitization, or after 11 applications there is no longer a difference in the irritation of the neck among the 3 groups of guinea pigs.

Two weeks later, macroscopic examination of the patch test on the flank shows an erythema and a tumefaction with no observable difference among the 3 groups. Thus, the resensitization has apparently succeeded.

On the other hand, histological examination reveals a definite difference between the first group, which was desensitized and resensitized, and the other 2 groups.

In the first group, there are practically no epidermal lesions and there is a diffuse infiltrate (Fig. 36); whereas in the other 2 groups, the epidermal and dermal lesions are similar to those usually seen in eczema due to DNCB (Fig. 37). Hence, the macroscopic and microscopic tests do not correspond.

This discrepancy has also been noted in another type of experiment on the sensitization to DNCB of leucopenic guinea pigs (RX, cytoxan). The patch test is positive, while histological examination shows neither spongiosis nor exocytosis and the perivascular infiltrate is minimal (MAGUIRE and MAIBACH, 1961; MAIBACH and MAGUIRE, 1963; MAGUIRE and MAIBACH, 1963; ZAGULA and coll., 1963).

Thus the results of the histological examination do not coincide with the aspect of the macroscopic lesions. However, we should keep in mind that, when we judge a macroscopic reaction, we appreciate only the dermal reactions; while with a histological examination, the degree of the epidermal lesions and the nature of the infiltrate are also realized.

*Thus from a macroscopic point of view, we can say that we have been able to resensitize guinea pigs, but we have not been able to demonstrate this resensitization histologically.*

*The resensitization of desensitized guinea pigs does not relegate these guinea pigs to their former state, but puts them into a new state which should be thoroughly investigated.*

## Remark

We cannot discuss desensitization without mentioning 3 more phenomena which in our opinion should be separated, at least today.

5 Hunziker, Experimental Studies

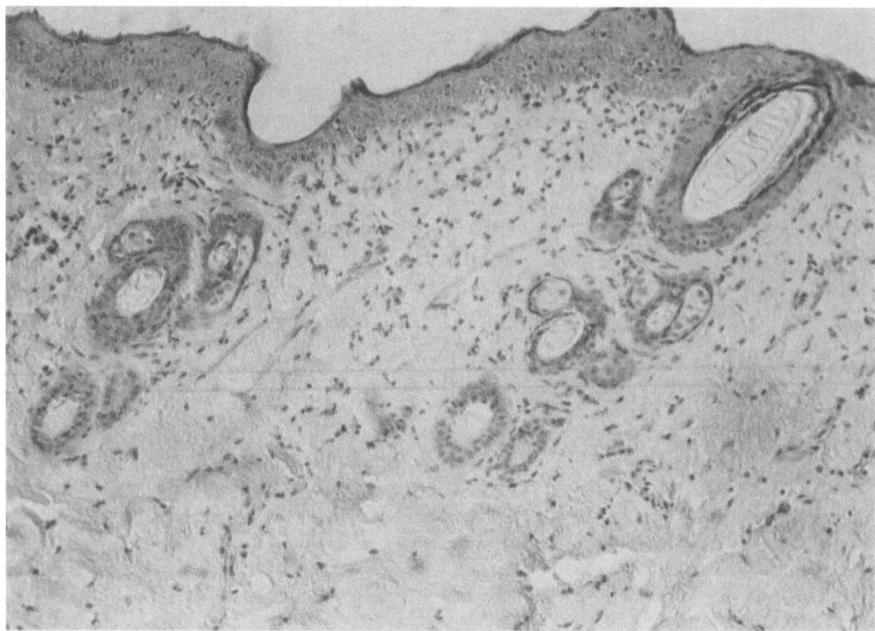

Fig. 36. Guinea pig from group one (desensitized) which was resensitized. Flank 14 hours after a single application of 0.1% DNCB in acetone: minimal epidermal alterations

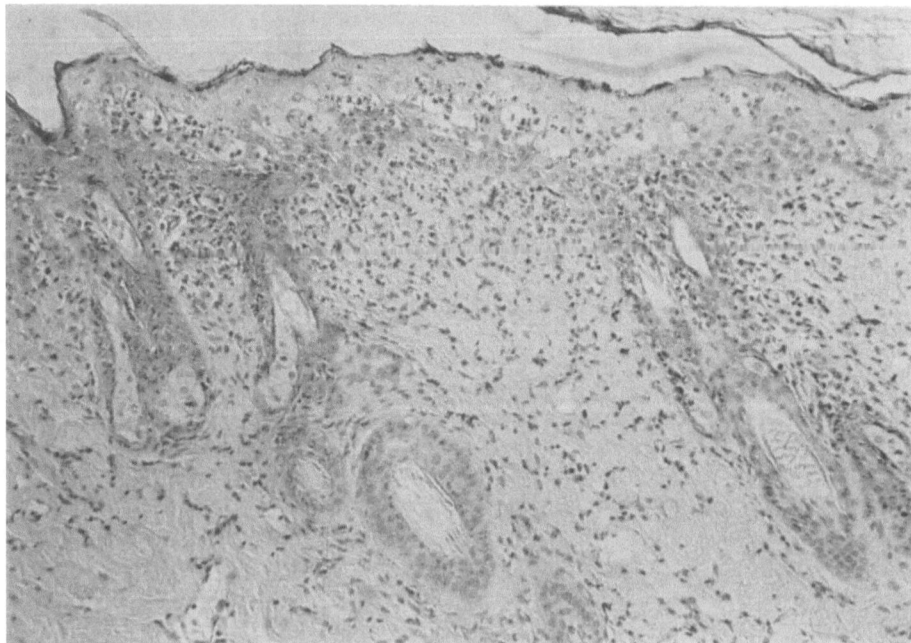

Fig. 37. Guinea pig from group two which was again sensitized. Flank 14 hours after a single application of 0.1% DNCB in acetone: epidermal alterations and infiltrate

1) *The triggering of a positive test can be rendered impossible by an injection of the allergen into the sensitized animal.*

A "tolerance" can be induced in guinea pigs already sensitized. It can be done in sensitized animals (according to the method of W. FREI, 1928), by injecting neo-arsphenamine* intravenously and 6 hours later intradermally (FREY and coll., 1964).

In animals sensitized to DNCB, an intravenous injection of a sodic solution of dinitro-benzensulfonic acid induces a complete inhibition of the contact reaction, but this is only temporary. Contrary to the injection of the dinitrobenzene sulfonic acid which produces a complete but temporary tolerance, the injection of simple DNCB does not induce a complete inhibition but only a decrease in the reaction (FREY and GELEIK, 1962; FREY and DE WECK, 1964 and DE WECK and FREY, 1966).

FREY and GELEIK (1962) after having compared their results and ours on desensitization, wrote: „Wir können nicht beantworten, ob es sich in beiden Fällen um das gleiche Phänomen handelt." We think that we can still state this today.

2) *Sensitization can be made impossible by previously administering the allergen to guinea pigs by injection, orally, or by epicutaneous application.*

W. FREI (1928) and SULZBERGER (1929) discovered that the guinea pigs became resistant to sensitization if a solution of neoarsphenamine was previously injected into the vein. This phenomenon is not regularly reproducible; it could not be repeated by CHASE (1949), but was reproduced by FREY and coll. (1964).

CHASE (1946) was able to demonstrate that previous oral administration of DNCB to guinea pigs can prevent sensitization to DNCB; this phenomenon is specific (experiments with 2 eczematogens: DNCB and orthochlorobenzacyl). Nevertheless, the oral antigen doses cannot induce a desensitization in the already sensitized animal (CHASE and BATTISTO, 1959).

Experiments of the same nature as CHASE's (1946) but with dinitrophenyl have been made by COE and SALVIN (1963) with negative results.

LOWNEY (1965) demonstrated a state of "immunological unresponsiveness to contact sensitizers" in guinea pigs, by application of PNDMA, DNCB and chlorprothixen.

Several attempts have been made to transpose CHASE's technique in order to be able to employ it to man to prevent sensitization, but without great success up until now (WHITE and BAER, 1950; GROLNICK, 1951; KLIGMAN, 1958).

The phenomenon described by FREI (1928), SULZBERGER (1929) and CHASE (1946) called: "Immunologic unresponsiveness to allergenic chemicals" by CHASE and BATTISTO (1959) is probably related to "immunoparalysis" discovered by FELTON (1949) and FELTON and coll. (1955)** and to the phenomenon of immunotolerance BILLINGHAM, BRENT) and MEDAWAR 1953, BRENT (1959)*** as well as to the tolerance to proteinic antigens (WEIGLE, 1959).

3) *By administering the allergen orally to pregnant guinea pigs one diminishes the capacity of sensitization either of the offspring.*

BAER and coll. (1958) and HARBER and coll. (1961 and 1962), relying on the experiments of BILLINGHAM, BRENT and MEDAWAR (1953), showed an acquired tolerance to DNCB, by a prenatal exposure to the allergen: pregnant guinea pigs are fed with DNCB; their offsping, when adult, are sensitized to DNCB and compared to controls; their capacity

---

\* Neoarsphenamine produces a tuberculin-type reaction.

\*\* Injections of the polysaccharides of pneumococci in appropriate doses which remain in the tissues for a long time and prevent the animal from responding to the immunization procedure.

\*\*\* Production of tolerance in the homografts by the cells or cellular extracts.

to become sensitive is decreased. This effect is specific since it does not at all influence the capacity of these guinea pigs to be sensitized with citraconic anhydride.

ROSENTHAL and BAER (1963) then asked themselves if one could induce a tolerance in new-born guinea pigs by injecting (intraperitoneally or subcutaneously) DNCB into these animals; it seems that in this case there was a tendency towards decreasing the capacity for sensitization when the guinea pigs became adult.

In opposition, SCHIMPF and FILIP (1965) were not able to succeed in inducing an immunotolerance (allergo-tolerance) in the same guinea pigs (now adult) by a paranatal treatment consisting of epicutaneous applications of DNCB.

We should like to recall that it is quite possible to sensitize the newborn guinea pigs with DNCB by epicutaneous applications (see p. 21). This fact explains the almost negative experiments of ROSENTHAL and BAER (1963) and the completely negative ones of SCHIMPF and FILIP (1965) in their attempt to obtain a "tolerance" in the adult guinea pigs by a paranatal treatment.

To summarize, *we obtain a general desensitization by repeated applications of the eczematogen to a very small surface (nipple) of the guinea pig sensitized to DNCB and to PNDMA.*

We should like to compare this general desensitization with three phenomena described in the literature:

1) The triggering of a positive test can be rendered impossible by injecting the allergen into the already sensitized animal.

2) The sensitization can be made impossible by previously administering the allergen to guinea pigs.

3) The capacity of sensitization, of the offspring when adult, can be diminished by administering the allergen orally to pregnant guinea pigs.

However, the facts do not yet enable us to decide whether these last three phenomena are related to each other and we cannot as yet state whether these phenomena are related to the general desensitization by repeated applications in guinea pig eczema to DNCB.


# VIII. Discussion of Some Questions

It is clear, as we have already stated, that experimental studies concerning contact eczema are for various reasons better done on animals than on man. For this reason, guinea pigs have been particularly used by most of the authors studying eczema. The experimental methods used are often different and it seems, at first, that a comparison is difficult. "Mais l'analyse même de ces variations fait apparaître le caractère essentiellement *individuel* de l'eczéma experimental, qui rejoint, par là même, l'eczéma humain" (DE GRACIANSKY and coll., 1960).

1) *Can parallels be drawn between experimental observations in guinea pigs and experimental and clinical observations in man?*

DNCB and PNDMA, CA and PA are substances rarely used in everyday life; that is why an eczema due to these products is rare in man. We know, however, that DNCB and PNDMA have been used by several authors in an attempt to sensitize man experimentally.

*The eczematogenic power of chemically simple substances often appears to be similar in guinea pigs and in man.* However, there are certain differences. This is why it is not always possible to test the capacity of an eczematogen on the guinea pig and then extrapolate the results to man.

2) *Is there a correlation between the lesions revealed by the histological examination of the reaction (patch test) in the guinea pig and those shown in the histological examination of the reaction in man?*

The interpretation of the reaction on the flank of the guinea pig is often difficult because of the thinness of the epidermis. For this reason W. JADASSOHN, BUJARD and BRUN (1955) used the nipple of the guinea pig whose epidermis is thicker, and where the *eczema is practically identical to the eczema of man.*

The resemblance of the eczematous reactions of the guinea pig and man in the histological examination has also been remarked by several authors: BLOCH and STEINER-WOURLISCH (1930) with primine eczema, BRUNSTING and BAILEY (1935) with ragweed, BAER, ROSENTHAL and SIMS (1956), GRIMMER and SPIER (1961) with DNCB etc.

GINGSBERG and coll. (1937 and 1939) report that, though there are certain resemblances in the histological examination in the guinea pig and man, there are also differences which cannot be attributed only to a difference in the degree of sensitization. They did not observe spongiosis in guinea pigs sensitized to DNCB; whereas, we noted it very frequently, even on the nonacanthosed flank.

A comparison between the development of the eczematous lesions in the guinea pig and man according to the observations of W. EPSTEIN and KLIGMAN (1959) shows good analogies (PILLSBURY, discussion of BAER, ROSENTHAL and SIMS, 1956).

For MIESCHER (1962) the allergic reaction of contact in the animal corresponds more or less with the allergic reaction in man: beginning of dermal inflammation, invasion of monocytes in the oedematous and dissociated epidermis. The differences observed by MIESCHER are the following:

a) The infiltration of the dermis is diffuse in the guinea pig; whereas, in man it is perivascular.

— We have also observed in the guinea pig not only a diffuse infiltration with numerous preparations, but also perivascular infiltrates —.

b) The invasion of the epidermis by spongiosis formation is not localized in the guinea pig as much as in man.

— However, we have often observed localized spongiosis on the nipple as well as on the flank of the guinea pig sensitized with DNCB and other simple chemical substances (N. HUNZIKER and coll., 1964) — (Fig. 6, 9, 10).

c) The vesicles originating from the spongiosis, whose framework of epithelial cells is still well conserved, exists only in a rudimentary form in the guinea pig.

— On the other hand, we have often observed vesicles in guinea pigs sensitized to DNCB whose framework of epithelial cells is still well conserved — (Fig. 13).

Let us recall that we observed lesions resembling those of "toxic epidermal necrolysis" (Lyell's disease) in guinea pigs sensitized with PA and tested with PA.

This is also an analogy between man and the guinea pig.

Although there are differences between the eczema of the guinea pig and that of man, it must however be stated that they are not very pronounced and need not essentially be due to the fact that there are species differences between the guinea pig and man, they may also be due to differences in the eczematogens, for we have shown that various eczematogens in the same animal show very pronounced variations (see p. 48).

3) *Are the reactions in eczema first epidermal and then subsequently dermal; the contrary or simultaneous?*

HAXTHAUSEN (1949) had already formulated the following: "It is to be emphasized, however, that in a great many cases of allergic eczema, probably in the majority of them, there is besides the epidermal hypersensitiveness, also a hypersensitiveness localized in the corium".

ST. EPSTEIN (1952, 1956 and 1958) also speaks of 2 shock organs, the epidermis and the dermis. It would seem that the epidermis and the dermis are more or less independant from each other in eczema. Both can react, or only one.

In our experiments, we observed that the epidermal lesions are almost always related to the dermal lesions. Already approximately two and a half hours after a test with DNCB in the guinea pig sensitized to this substance, one sees slight spongiosis, which may or may not be invaded by lymphocytes, and a dermal infiltrate. In contrast, in our experiments on eczema with potassium bichromate, there are practically no epidermal lesions, but an intense infiltrate in the dermis (p. 43).

This shows different lesions produced by two eczematogenes; but one must add that the method of sensitization was also different. Epicutaneous application was performed for DNCB and injections were used with Freund's adjuvant for chromium.

In our clinic, DE WECK and BRUN (1956) observed, by means of passive transfer of donor animals sensitized with picryl chloride (according to CHASE's method by injection of Freund's adjuvant), a histological picture in the receiver animals showing dermal but no epidermal lesions. GRIMMER and SPIER (1961) however, succeeded in transfering sensitivity to DNCB from animals sensitized to DNCB (by epicutaneous application). This was confirmed by a histological examination which showed both dermal and epidermal lesions. Nevertheless, their method of sensitization was different from the one used by DE WECK and BRUN.

GRIMMER (1961) pointed out that if sensitization is carried out through the lymph nodes or the spleen, only a dermal reaction is obtained. Recently this has also been emphasized by KLASCHKA (1966). The histological picture is different according to the means of sensitization (DNCB): solely dermal lesions (sensitization by intravenous and intraperitoneal injections) or combined dermal and epidermal lesions (epicutaneous applications or intracutaneous injections).

Solely epidermal lesions were pointed out by SATO (1961) in guinea pigs sensitized with phenyl-$\beta$-naphthylamine.

In man, ST. EPSTEIN (1952, 1956 and 1958) demonstrated tests with neomycin and nickel methiolate, which showed only dermal reactions. For W. EPSTEIN and KLIGMAN (1959), as for others, the dermal reactions precede the epidermal lesions, the lymphocytes play an essential part in the eczematous reaction.

Our experiments and those of SPIER and his school carried out on guinea pigs indicate, like those of ST. EPSTEIN carried out in man, that the skin has an allergic „double manteau", one in the epidermis and the other in the dermis.

We should also like to add that the primordial "vésiculette" described by CIVATTE (1925, 1950 and 1954) in man, as a first sign of eczema, is very controversial in its significance (POLAK and MOM, 1949; MIESCHER, 1952; BANDMANN, 1960 etc.).

In guinea pigs sensitized with DNCB we have rarely noticed such "vésiculettes" in adults, but often in the new-born. Nevertheless, they were always accompanied by other signs of eczema.

In man, for CIVATTE (1925) the "vésiculette" is a sign of eczema; for MIESCHER (1952) it is a sign of a toxic reaction. For us, in the guinea pig, it is a sign of eczema but we are unable to say whether it is a primordial sign, because we have always found "vésiculettes" and spongiosis simultaneously.

*To summarize*, it can be said that one finds pure dermal reactions (sensitization with potassium bichromate, passive transfer according to DE WECK and BRUN (1956) etc.) and pure epidermal reactions (SATO, 1961). It seems that epidermal and dermal reactions are more or less independent of one another.

4) *Is there a relation between urticarial and eczematous reactions?*

It seems that one has to differentiate between the urticarial reaction (immediate reaction) and the eczematous reactions (delayed reactions) (W. JADASSOHN, 1932).

The combination of immediate and delayed reactions is rare in man (e.g.: ammonium persulfate) (CALNAN and SHUSTER, 1963; BRUN, W. JADASSOHN and PAILLARD, 1966; and W. JADASSOHN, 1966).

In the guinea pig this combination was obtained by several authors (DNCB, picryl chloride, etc.): immediate anaphylactic* reaction and delayed reaction together.

In guinea pigs sensitized with CA, we obtained after a test an immediate urticarial reaction which evolved after 7—8 hours into a delayed eczematous reaction.

Urticarial sensitivity can be transfered by the serum, whereas it is admitted that eczematous sensitivity could only be transfered by cells. With CA we succeeded in transfering by the serum, according to a technique similar to that of PRAUSNITZ-KÜSTNER, of not only an urticarial reaction but also a reaction of the delayed type lasting for 12 to 24 hours. The infiltrate, which is composed mainly of eosinophils, increases considerably during the hours that follow the test and the epidermis shows rough outlines of spongiosis.

In sensitization with CA it is a matter of combined urticarial and eczematous reactions and they cannot be clearly distinguished from each other.

Furthermore, another anhydride, PA, does not seem to provoke a clear urticarial reaction and it cannot be transfered by the serum. The infiltrate contains eosinophils which are less numerous, however, than in the reaction with CA.

5) *Do certain eczema-producing substances have a more general effect than the simple effect localized on the skin?*

a) According to our observations we can say that one sensitization does not protect against another sensitization with a different eczematogen. Therefore we can say, as WEDROFF and DOLGOFF (1935) wrote concerning man: „Bei unseren Experimenten wirkte die experimentelle Sensibilisierung gegen Dinitrochlorobenzol nicht auf den Verlauf einer früher bestandenen Allergie gegen einen anderen Stoff ein".

We cannot say that a first sensitization favors a new sensitization.

b) We observed two hematological variations in relation to sensitization in the guinea pig:

---

* Demonstrated by the anaphylactic shock, by the Schulz-Dale method and by the passive transfer method of PRAUSNITZ-KÜSTNER modified by CHASE.

α) *A blood eosinophilia* persists following sensitization in animals sensitized with CA (see p. 44) and varies in those sensitized with PA. In the eczematous sensitization we did not observe a continuous parallel between this blood eosinophilia and the accumulation of eosinophils in the patch test. Their number in patch tests is not influenced by corticosteroids, unlike blood eosinophilia. In passive transfer by serum the eosinophils are numerous in the patch test, whereas their number is normal in the recipient blood.

β) *A blood basophilia* appears toward the end of sensitization then it disappears in a few days (see p. 45). The significance of this basophilia is not yet clear to us; but this phenomenon also exists in different types of sensitization (foreign proteins). Thus, it links the sensitization due to chemically simple substances to the larger group of sensitization of the delayed type.

6) *Unspecific dispositions to eczema.*

We ascertained that it may be difficult to sensitize certain guinea pigs, even with strong eczema-producing substances (e.g. DNCB). This was also demonstrated by CHASE (1941). As in man, there are individual differences, an „Ekzembereitschaft" as BLOCH (1923) called it, which means that certain individuals would have a greater tendency to sensitization than others.

In a group of guinea pigs which we were not able to sensitize successfully with a first eczema-producing substance (PNDMA), we noted subsequently that successful sensitization was difficult even with a strong eczema-producing substance such as DNCB. This is in opposition to another group of guinea pigs which had been successfully sensitized with PNDMA (see p. 44). There is not only a specific „Ekzembereitschaft" but also an unspecific one.

# IX. Conclusion

When we work on the experimental eczema of the guinea pig, we often wonder if these experiments are meaningful.

We have two reasons for answering in the affirmative:

a) The eczema of the guinea pig is in itself a very interesting and theoretically important phenomenon.

b) The relationships between the eczema of the guinea pig and that of man are very important, as a great number of facts reported in this study show. These relationships are greater than certain authors are willing to admit and we feel that this must be emphasized.

Since the first articles of BLOCH and STEINER-WOURLISCH and of W. JADASSOHN, which date from 1930, many investigators have studied the problem of eczema in the guinea pig and there is still much to do, in spite of the number of experiments which have been carried out on this animal. One might almost say that it is still a matter of preliminary studies, because most experiments were carried out on one kind of animal only, the guinea pig, and mostly with the same eczematogen: DNCB. Although we only experimented on the guinea pig for the time being, our research shows clearly that other eczema-producing substances sometimes give quite different results.

We hope that this study showed that it is worth while, or rather indispensable, to continue further investigations and to gather facts.

It is possible that later, through other observations, and other experiments, one will obtain results not only of theoretical importance, but also of practical value.

# Literature

## Introduction. Material and Technique

BLOCH, BR.: Ekzem und Allergie. C.R. séances VIII congrès intern. de derm. et syph. 1930. p. 99—107. Copenhagen: Engelsen and Schroder 1931.

—, u. A. STEINER-WOURLISCH: Die willkürliche Erzeugung der Primelüberempfindlichkeit beim Menschen und ihre Bedeutung für das Idiosynkrasieproblem. Arch. Derm. Syph. (Berl.) **152**, 283—303 (1926).

— — Die Sensibilisierung des Meerschweinchens gegen Primeln. Arch. Derm. Syph. (Berl.) **162**, 349—378 (1930).

BUJARD, E.: La phosphatase alcaline dans la tétine de cobaye et le „Nipple-test". Dermatologica (Basel) **115**, 181—185 (1957).

CASH, J. T.: The dermatitis produced by east Indian satinwood ("chloroxylon swietenia"). Brit. med. J. **1911** II, 784—790.

CHASE, M. W.: Inheritance in guinea pigs of the susceptibility to skin sensitization with simple chemical compounds. J. exp. Med. **73**, 711—726 (1941).

— Experimental sensitization with particular reference to picryl chloride. Int. Arch. Allergy **5**, 163—191 (1954).

DOERR, R.: Allergische Phänomene. Handb. norm. u. pathol. Physiologie, Bd. XIII, S. 650—812. Berlin: Springer 1929.

FISHER, J. P., and H. A. COOKE: Experimental toxic and allergic contact dermatitis. I. A chemical study of histamine content. J. Allergy **29**, 396—410 (1958).

FREI, W.: Ein Versuch zwischen überlebender Haut von Tuberkulin-Trichophytin- oder Quecksilber-Überempfindlichkeit und den zugehörigen „Antigenen" außerhalb des Körpers spezifische Reaktionen nachzuweisen. Klin. Wschr. **7**, 457—458 (1928).

— Über willkürliche Sensibilisierung gegen chemisch-definierte Substanzen. II. Mitt. Untersuchungen mit Neosalvarsan am Tier (Salvarsanexantheme beim Tier). Klin. Wschr. **7**, 1026—1031 (1928).

GOMORI, G.: The distribution of phosphatase in normal organs and tissues. J. cell. comp. Physiol. **17**, 71—80 (1941).

HUNZIKER, N.: Experimental eczema. 21st comm. Alkaline phosphatase in eczema of the guinea pig treated with dinitrochlorobenzene. Dermatologica (Basel) **129**, 359—369 (1964).

JADASSOHN, J.: Zur Kenntnis der medikamentösen Dermatosen. Verh. dtsch. Derm. Gesell. **5**, 103 (1896). V. Kongress.

— Aetiologie und Pathogenese der Ekzem. VIII congrès intern. de derm. et syph. 1930, p. 64—98. Copenhagen: Engelsen & Schroder 1931.

JADASSOHN, W.: Sensibilisierung der Haut des Meerschweinchens auf Phenylhydrazin. Klin. Wschr. **9**, 551—552 (1930).

— Die Immunbiologie der Haut. Handb. Haut. u. Geschlechtskr. II, S. 355—478. Berlin: Springer 1932.

— E. BUJARD, and R. BRUN: The experimental eczema of the guinea-pig nipple. J. invest. Derm. **24**, 247—253 (1955).

LANDSTEINER, K., and M. W. CHASE: Studies on the sensitization of animals with simple chemical compounds. VII. Skin sensitization by intraperitoneal injections. J. exp. Med. **71**, 237—245 (1940).

— — Studies on the sensitization of animals with simple chemical compounds. IX. Skin sensitization induced by injection of conjugates. J. exp. Med. **73**, 431—438 (1941).

LANDSTEINER, K., and J. JACOBS: Studies on the sensitization of animals with simple
    chemical compounds. J. exp. Med. **61**, 643—656 (1935).
LOW, R. C.: Skin-sensitiveness to non-bacterial proteins and toxins. Brit. J. Derm. Syph.
    **36**, 292—312 (1924).
MACHER, E., u. W. SENNLAUB: Über die Sensibilisierbarkeit von Meerschweinchen durch
    Applikation von Dinitrochlorbenzol in Lymphknoten oder Milz. Dermatologica
    (Basel) **126**, 207—222 (1963).
MAYER, R. L., u. M. B. SULZBERGER: Zur Frage der jahreszeitlichen Schwankungen der
    Krankheiten. Der Einfluß der Kost auf experimentelle Sensibilisierungen. Arch. Derm.
    Syph. (Berl.) **163**, 245—262 (1931).
MIESCHER, G.: Ekzem. Histopathologie, Morphologie, Nosologie. Handb. Haut.- u.
    Geschlechtskr. Erg. II/1, S. 1—112. Berlin-Göttingen-Heidelberg: Springer 1962.
NESTLER: see LOW.
NILZEN, A.: Some endocrine aspects of skin sensitization and primary irritation. 1. Obser-
    vations of the influence of cortisone on cutaneous irritation and sensitization induced by
    various chemical compounds. J. invest. Derm. **18**, 7—35 (1952).
ROCKWELL, E.: Study of several factors influencing contact irritation and sensitization.
    J. invest. Derm. **24**, 35—49 (1955).
SPIER, H. W.: Allergie der Haut. In: GOTTRON, H. A., u. W. SCHÖNFELD: Dermatologie
    und Venerologie. Bd. I/1, S. 613—704. Stuttgart: Thieme 1961.
STORCK, H.: Das experimentelle Ekzem. Handb. Haut- u. Geschlechtskr. Erg. II/1,
    S. 113—211. Berlin-Göttingen-Heidelberg: Springer 1962.
SULZBERGER, M. B.: Hypersensitiveness to Arsphenamine in guinea-pigs. 1. Experiments in
    prevention and in desensitization. Arch. Derm. (Chic.) **20**, 669—697 (1929).
—, et V. H. WITTEN: Quelques caractéristiques du processus de l'eczéma de type contact
    chez l'homme. Immunologie et morphologie comparative. In: CHARPY, J. (Ed.): Le
    mécanisme physio-pathologique de l'eczéma, p. 115—129. Paris: Masson 1954.
WALTHARD, B.: Die Erzeugung experimenteller Nickelidiosynkrasie bei Laboratoriums-
    tieren. Schweiz. med. Wschr. **56**, 603—604 (1926).
WEDROFF, N. S., u. A. P. DOLGOFF: Ein Beitrag zur Frage der Reaktionsfähigkeit der Haut
    bei „chemischen" Ekzemen, auf Grund von Untersuchungen bei ekzemkranken und
    gesunden Personen mittels der obligaten chemischen Reizmittel. Arch. Derm. Syph.
    (Berl.) **171**, 641—646 (1935).

## I. Experimentation with Dinitrochlorobenzene (DNCB)

ANKE, H.: Zur Frage der Ausbreitung der Sensibilisierung in der Epidermis. Derm. Wschr.
    **109**, 1263—1264 (1939).
BAER, R. L., St. A. ROSENTHAL, and CH. SIMS: Contact dermatitis with spongiosis and
    intraepidermal vesiculation in the acanthotic skin of guinea pigs. J. invest. Derm. **27**,
    249—258 (1956).
— — — The allergic eczemalike reaction and the primary irritant reaction. Arch. Derm.
    (Chic.) **76**, 549—560 (1957).
— CH. F. SIMS u. St. A. ROSENTHAL: Einige Wirkungen der Acanthose auf die Reaktions-
    fähigkeit der Meerschweinchenhaut. Hautarzt **10**, 199—202 (1959).
BALDRIDGE, G. D., and A. M. KLIGMAN: Preliminary and short report. The effect of corti-
    sone on experimentally induced contact dermatitis. J. invest. Derm. **17**, 257—259 (1951).
BANDMANN, H. J.: Beitrag zur Histopathologie allergischer epicutaner Testreaktionen.
    Teil I. Das allergische Kontaktekzem und die epicutane Läppchenprobe sowie bisher
    durchgeführte Untersuchungen zur Histopathologie der durch den Läppchentest
    hervorgerufenen Reaktionen. Hautarzt **11**, 258—262 (1960).
— Beitrag zur Histopathologie allergischer epicutaner Testreaktionen. Teil II. Allgemeine
    Anordnung der eigenen Versuche, spezielle Versuchsanordnung und Ergebnisse der
    Untersuchungen. Hautarzt **11**, 310—318 (1960).
— Beitrag zur Histopathologie allergischer epicutaner Testreaktionen. Teil III. Ergebnisse
    der Untersuchungen. Hautarzt **11**, 355—363 (1960).

Bujard, E., R. Brun et W. Jadassohn: Expériences sur l'acanthose chez le cobaye. Dermatologica (Basel) **114**, 171—177 (1957).

— W. Jadassohn, R. Brun et R. Paillard: Des effets, mito-excitateur ou nécrosant, de quelques substances sensibilisantes. Acta allerg. (Kbh.) **6**, 161—167 (1953).

Charov, J., A. Stahl et P. Y. Castelain: Etude histologique et chronologique de la constitution de la lésion d'eczéma. In: Charpy, J. (Ed.): Le mécanisme physio-pathologique de l'eczéma, p. 39—55. Paris: Masson 1954.

Chase, M. W.: Inheritence in guinea pigs in the susceptibility to skin sensitization with simple chemical compounds. J. exp. Med. **73**, 711—726 (1941).

— Experimental sensitization with particular reference to picryl chloride. Int. Arch. Allergy **5**, 163—191 (1954).

—, and J. R. Battisto: The duration of dermal sensitization following cellular transfer in guinea pigs. J. Allergy **26**, 83 (1955).

Civatte, A.: Eczéma et eczématides. Bull. Soc. franç. Derm. Syph. **32**, 134—147 (1925).

— Anatomie pathologique de l'eczéma. Bull. Soc. franç. Derm. Syph. **57**, 13—32 (1950).

— La formule histologique de l'eczéma. In: Charpy, J. (Ed.): Le mécanisme physio-pathologique de l'eczéma, p. 56—60. Paris: Masson: 1954.

Epstein, W. L., and A. M. Kligman: Some factors affecting the reaction of allergic contact dermatitis. J. invest. Derm. **33**, 231—243 (1959).

Fisher, J. P., and H. A. Cooke: Experimental toxic and allergic contact dermatitis. 1. A chemical study of histamine content. J. Allergy **29**, 396—410 (1958).

— — Experimental toxic and allergic contact dermatitis. II. A histopathologic study. J. Allergy **29**, 411—428 (1958).

—, and D. Glick: Histochemistry. XIX. Localization of alkaline phosphatase in normal and pathological human skin. Proc. Soc. exp. Biol. (N.Y.) **66**, 14—18 (1947).

Freund, J.: The sensitiveness of tuberculous guinea pigs one month old to the toxicity of tuberculin. J. Immunology **17**, 465—471 (1929).

Frey, J. R.: Quantitative Untersuchungen bei epidermaler Sensibilisierung von Meerschweinchen mit Dinitrochlorbenzol. Dermatologica (Basel) **102**, 1—11 (1951).

—, and A. Studer: Cortison und experimentelles Kontaktekzem mit Dinitrochlorbenzol am Meerschweinchen. Dermatologica (Basel) **103**, 65—74 (1951).

—, u. P. Wenk: Experimentelle Untersuchungen zur Pathogenese des Kontaktekzems. Dermatologica (Basel) **112**, 265—305 (1956).

Frumess, G. M.: Allergic reaction to dinitrophenol. J. Amer. med. Ass. **102**, 1219—1220 (1934).

Gans, O., u. G. K. Steigleder: Histologie der Hautkrankheiten. 2. Aufl., S. 295. Berlin-Göttingen-Heidelberg: Springer 1955.

Gaudin, P.: Acanthose par chrysarobine, vaseline et frottement. Dermatologica (Basel) **97**, 208—215 (1948).

Gentele, H., and H. J. Holmgreen: Investigation of a possible inhibitor action of Heparin upon certain experimental allergic dermatoses. Acta derm.-venereol. (Stockh.) **31**, 322—330 (1951).

Gershbein, L. L., B. K. Krotoszynsky, M. J. Brunner, and A. J. Domnas: Respiratory activity and nucleic acid content of some tissues from guinea pigs sensitized to 2,4-dinitrochlorobenzene. J. invest. Derm. **27**, 73—76 (1956).

Ginsberg, J., F. Becker, and W. Becker: Sensitization of guinea pigs to poison Ivy. Arch. Derm. Syph. (Chic.) **36**, 1165—1170 (1937).

—, C. D. Stewart, and S. W. Becker: Cutaneous sensitization studies. II. Cross and microscopic changes in ragweed and 2-4 dinitrochlorobenzene sensitization of guinea pigs, and in poison Ivy sensitization of human beings. J. invest. Derm. **2**, 81—92 (1939).

Golay, M., et R. Brun: De l'eczéma expérimental. 5ème comm. La réaction eczémateuse déclenchée au cours de la sensibilisation du cobaye. Dermatologica (Basel) **116**, 408—412 (1958).

— — De l'eczéma expérimental. 6ème comm. Eczéma et acanthose provoqués chez le cobaye par applications répétées de dinitrochlorobenzène. Dermatologica (Basel) **116**, 412—415 (1958).

Götz, H., u. J. Schulz: Zur Frage der Beziehungen zwischen Pilzinfektion und epidermaler Sensibilisierung gegen Dinitrochlorbenzol beim Meerschweinchen. Arch. klin. exp. Derm. **203**, 577—581 (1956).

De Graciansky, P., L. Israel et J. Cohen-Solal: L'eczéma du cobaye au dinitrochlorobenzène. Etude de l'influence de certains médicaments neurotropes. Sem. Hôp. Paris **36**, 1451/S P 303—1457/SP 309 (1960).

— — — L'eczéma du cobaye au dinitrochlorobenzène. II. Détermination quantitative de la sensibilité. Notion de seuil. Individualité des cobayes. Sem. Hôp. Paris **36**, 1458/SP 310 —1460/SP 312 (1960).

— — — L'eczéma du cobaye au dinitrochlorobenzène. III. Influence de certains médicaments neurotropes sur l'eczéma expérimental du cobaye. Sem. Hôp. Paris **36**, 1461/SP 313—1464/SP 316 (1960).

Grimmer, H.: Die Histologie der epicutanen Testreaktion bei nicht epicutaner („extracutaner") Sensibilisierung mit 2,4-Dinitrochlorbenzol. Arch. klin. exp. Derm. **214**, 105—113 (1961).

—, u. H. W. Spier: Histologische Verifizierung der passiven Übertragbarkeit des tierexperimentellen Kontaktekzems. Zur Interferenz toxischer und allergischer Epicutanreaktionen. Hautarzt **12**, 300—306 (1961).

Halter, K.: Zur Pathogenese des Ekzems. Arch. Derm. Syph. (Berl.) **181**, 592—719 (1941).

Haxthausen, H.: Some problems concerning the pathogenesis of allergic eczemas, elucidated by experiments on sensitization with dinitrochlorobenzene. Acta derm.-venereol. (Stockh.) **20**, 257—272 (1939).

— Further experiments on sensitization of the skin with dinitrochlorobenzene. Acta derm. venereol. (Stockh.) **21**, 158—165 (1940).

— The pathogenesis of allergic eczema elucidated by transplantation. Experiments on identical twins. Acta derm. venereol. (Stockh.) **23**, 438—457 (1943).

— Passive transmission of dinitrochlorbenzene allergy with white blood cells from sensitized guinea pigs. Acta derm. venereol. (Stockh.) **31**, 659—665 (1951).

Hösli, F.: Untersuchungen über den Einfluß der Höhenklimas auf allergische Reaktionsvorgänge. Dermatologica (Basel) **96**, 151—162 (1948).

Hüllstrung, H., u. K. Hack: Experimentelle Sensibilisierung und deren Beeinflussung durch Vitamine beim Meerschweinchen. Z. Immun-Forsch. **100**, 393—428 (1941).

Hunziker, N.: De l'eczéma expérimental. 14ème comm. Effet de multiples applications de dinitrochlorobenzène chez le cobaye. Dermatologica (Basel) **123**, 326—330 (1961).

—, et G. Schinas: Expériences sur cobayes nouveau-nés. Eczéma au dinitrochlorobenzène. 15ème comm. Dermatologica (Basel) **124**, 235—239 (1962).

Jadassohn, W.: Zur Wirkung von Chrysarobin auf die Haut. Schweiz. med. Wschr. **71**, 1143—1145 (1944).

—, E. Bujard, and R. Brun: The experimental eczema of the guinea-pig nipple. J. invest. Derm. **24**, 247—253 (1955).

—, E. Uehlinger u. H. E. Fierz: Über die Wirkung von weiblichen Sexualhormonen auf das Epithel der Meerschweinchenzitze. Schweiz. med. Wschr. **71**, 6—8 (1941).

Kalkoff, K. W.: Experimentelle Studien über den Vorgang der epidermalen Sensibilisierung. I. Mitt. Über die intraepidermale Ausbreitung der ekzematösen Sensibilisierung. Arch. Derm. Syph. (Berl.) **186**, 493—511 (1948).

Klaschka, F.: Tierexperimentelle Studien zum Kontaktekzem: I. Histologie der epicutanen 2,4-Dinitrochlorbenzol-Hautreaktion bei intraperitoneal nach Landsteiner sensibilisierten Meerschweinchen. Arch. klin. exp. Derm. **218**, 475—498 (1964).

— Tierexperimentelle Studien zum Kontaktekzem. II. Histologie der epicutanen 2,4-Dinitrochlorbenzol-Hautreaktion bei „cutan" und „extracutan" sensibilisierten Meerschweinchen. Arch. klin. exp. Derm. **224**, 216—234 (1966).

Kopf, A. W.: The distribution of alkaline phosphatase in normal and pathologic human skin. Arch. Derm. Syph. (Chic.) **75**, 1—37 (1957).

LANDSTEINER, K., and M. W. CHASE: Studies on the sensitization of animals with simple chemical compounds. IV. Anaphylaxis induced by picryl chloride and 2:4 dinitrochlorobenzene. J. exp. Med. **66**, 337—351 (1937).

— — Skin sensitization to simple chemical compounds by injections of conjugates. Proc. Soc. exp. Biol. (N.Y.) **44**, 559 (1940).

— — Studies on the sensitization of animals with simple chemical compounds. VII. Skin sensitization by intraperitoneal injections. J. exp. Med. **71**, 237—245 (1940).

— — Studies on the sensitization of animals with simple chemical compounds. IX. Skin sensitization induced by injection of conjugates. J. exp. Med. **73**, 431—438 (1941).

—, and J. JACOBS: Studies on the sensitization of animals with simple chemical compounds. J. exp. Med. **61**, 643—656 (1935).

— — Studies on the sensitization of animals with simple chemical compounds. J. exp. Med. **64**, 625—639 (1936).

LELOIR, H.: Anatomie pathologique de l'eczéma. Ann. Soc. franç. Derm. Syph. **1**, 465—477 (1890).

LEVER, W. F.: Histopathology of the skin. 3rd ed., p. 83. Philadelphia-Montreal: J. B. Lippincott 1961.

MACHER, E.: Die Reaktion der regionären Lymphknoten beim tierexperimentellen allergischen Kontaktekzem. I. Makroskopische Untersuchungen. Hautarzt **13**, 18—23 (1962).

—, u. W. SENNLAUB: Über die Sensibilisierbarkeit von Meerschweinchen durch Applikation von Dinitrochlorbenzol in Lymphknoten oder Milz. Dermatologica (Basel) **126**, 207—222 (1963).

MAIBACH, H. I., and H. C. MAGUIRE: Elicitation of delayed hypersensitivity (DNCB contact dermatitis) in markedly panleukopenic guinea pigs. J. invest. Derm. **41**, 123—127 (1963).

MAYER, R., u. M. B. SULZBERGER: Zur Frage der jahreszeitlichen Schwankungen der Krankheiten. Der Einfluß der Kost auf experimentelle Sensibilisierungen. Arch. Derm. Syph. (Berl.) **163**, 245—262 (1931).

MIESCHER, G.: Ist die Ausbreitung der allergischen Sensibilisierung ein Reflexvorgang? Schweiz. med. Wschr. **22**, 1360—1362 (1941).

— Zur Histologie der ekzematösen Kontaktreaktion. Dermatologica (Basel) **104**, 215—220 (1952).

— Ekzem. Histopathologie, Morphologie, Nosologie. In: Handb. Haut- u. Geschlechtskr., Erg. Bd. II/1, S. 1—111 (S. 40). Berlin-Göttingen- Heidelberg: Springer 1962.

NIEBAUER, G.: Nervensystem und allergisches Ekzem. Acta neuroveg. (Wien) suppl. VIII, 1—125 (S. 53) (1962).

NILZEN, A.: Some endocrine aspects of skin sensitization and primary irritation. I. Observations of the influence of cortisone on cutaneous irritation and sensitization induced by various chemical compounds. J. invest. Derm. **18**, 7—35 (1952).

— Some aspects of epidermal testing of guinea-pigs sensitized and not sensitized to 2,4-dinitrochlorobenzene. Acta derm.-venereol. (Stockh.) **32**, suppl. 29, 231—239 (1952).

—, and F. HUSSEY: Some endocrine aspects of skin sensitization and primary irritation. 4. Observations on the influence of thyroid deficiency and administration of thyroid hormone upon cutaneous irritation and sensitization reactions. J. invest. Derm. **22**, 503—514 (1954).

NISHIYAMA, S.: Capillardarstellung durch die alkalische Phosphatase-Färbung bei verschiedenen Dermatosen. II. Nicht infektiöse, entzündliche Dermatosen. Hautarzt **14**, 210—221 (1963).

PIRILÄ, V., and D. ERÄNKÖ: Distribution of histochemically demonstrable alkaline phosphatase in normal and pathological human skin. Acta path. microbiol. scand. **27**, 650—661 (1950).

POLAK, M., and A. M. MOM: Histopathology of experimental eczema (allergic contact-type eczematous dermatitis) in man. A study by the technics of silver impregnations of Rio Hortega with special reference to the early microscopic lesions. J. invest. Derm. **13**, 125—134 (1949).

RAAB, W.: Die Beeinflussung des experimentellen Dinitrochlorbenzolekzems des Meerschweinchens durch den Histaminliberator 48/80. Arch. klin. exp. Derm. 214, 307—318
(1962).

RAJKA, G., and S. HARD: Changes in guinea-pig skin resulting from repeated application of
dinitrochlorobenzene at the same site. Acta derm.-venereol. (Stockh.) 40, 64—73
(1960).

REBELLO, D. J. A., and R. R. SUSKIND: The effect of common contactants on cutaneous
reactivity to sensitizers. J. invest. Derm. 41, 67—80 (1963).

ROCKWELL, E. M.: Study of several factors influencing contact irritation and sensitization.
J. invest. Derm. 24, 35—49 (1955).

ROSENTHAL, ST. A., and R. L. BAER: Actively acquired tolerance to dinitrochlorobenzene.
Tests in newborn guinea pigs. J. invest. Derm. 41, 351—355 (1963).

ROSTENBERG, A.: Studies on the eczematous sensitization. I. The route by which the sensitization generalizes. J. invest. Derm. 8, 345—355 (1947).

—, and J. B. HAEBERLIN: Studies in eczematous sensitizations. III. The development in
species other than man or the guinea pig. J. invest. Derm. 15, 233—247 (1950).

SCHAAF, F., u. F. GROSS: Tierexperimentelle Untersuchungen mit Salben und Salben-
grundlagen. Dermatologica (Basel) 106, 357—378 (1953).

SCHEPANK, H.: Tierexperimentelle Untersuchung über die Beeinflussung der allergischen
Reaktionslage. Z. Haut- u. Geschl.-Kr. 18, 304—310 (1955).

SCHIMPF, A., u. G. FILIPP: Untersuchungen zur Frage der Immunotoleranz gegenüber
Dinitrochlorbenzol. Acta allerg. (Kbh.) 20, 187—196 (1965).

SCHNITZER, A.: Untersuchungen über den Ausbreitungsmechanismus der ekzematösen
Sensibilisierung. Dermatologica (Basel) 83, 70—79 (1941).

— Beitrag zur Frage des Mechanismus der Sensibilisierung. Dermatologica (Basel) 85,
339—347 (1942).

SCHREIBER, W., u. W. MÜLLER: Grundlegender Versuch zur Klärung der Frage, auf wel-
chem Wege sich die Sensibilisierung in der Epidermis ausbreitet. Derm. Wschr. 107,
1393—1395 (1938).

SCHULZ, K. H.: Chemische Struktur und allergene Wirkung. Aulendorf/Württ.: Cantor
1962.

SEEBERG, G.: Eczematogenous sensitization via the lymphatic glands as compared with
other routes. A study with 2:4 dinitrochlorbenzene. Acta derm.-venereol. (Kbh.) 31,
592—598 (1951).

SEEBOHM, P. M., M. H. TREMAINE, and W. S. JETER: The effect of cortisone and adreno-
corticotropic hormone on passively transferred delayed hypersensitivity to 2,4-dinitro-
chlorobenzene in guinea pigs. J. Immunology 73, 44—48 (1954).

SKOG, E.: Experimental studies on hypersensitivity to 2,4-dinitrochlorobenzene and tuberculin in animals. I. Passive transfer of hypersensitivity to 2,4-dinitrochlorobenzene.
Acta derm. venereol. (Kbh.) 35, 93—106 (1955).

SULZBERGER, M. B., and R. L. BAER: Sensitization to simple chemicals. III. Relationship
between chemical structure and properties, and sensitizing capacities in the production
of eczematous sensitivity in man. J. invest. Derm. 1, 45—58 (1938).

TZANCK, A., et G. R. MELKI: Contribution à l'étude histologique des tests épicutanés. In:
CHARPY, J. (Ed.): Le mécanisme physio-pathologique de l'eczéma, p. 109—112. Paris:
Masson 1954.

UNNA, P. G.: Lokale infektiöse Entzündungen. S. 162—344. Status spongioides. Ekzema
rubrum. Nichtinfektiöse Ekzemrecidive, S. 219—222. In: Histopathologie der Hautkrankheiten. Berlin: Hirchwald 1894.

VALTIS, J., et A. SAENZ: Sur la sensibilité générale à la tuberculine des jeunes cobayes.
C.R. Soc. Biol. (Paris) 99, 1562—1563 (1928).

DE WECK, A., et R. BRUN: De l'eczéma expérimental. 2ème comm. La sensibilisation du
cobaye au dinitrochlorobenzène et au chlorure de picryle. Dermatologica (Basel) 113,
335—368 (1956).

— — De l'eczéma expérimental. 3ème comm. A propos de l'effet protecteur de l'acanthose.
Dermatologica (Basel) 114, 91—101 (1957).

Wedroff, N.: Die Tropfmethode zur Prüfung der Überempfindlichkeit der Haut gegen chemische Stoffe. Arch. Derm. Syph. (Berl.) **167**, 224—232 (1932).

Wedroff, N. S., u. A. P. Dolgoff: Über die spezifische Sensibilität der Haut einfachen chemischen Stoffen gegenüber. Arch. Derm. Syph. (Berl.) **171**, 647—664 (1935).

Wolf-Eisner, A.: Dermatol. centr. **10**, 164 (1907).

Zeligman, I.: Experimental contact dermatitis. I. Dinitrochlorobenzene contact dermatitis in guinea pigs. J. invest. Derm. **22**, 109—120 (1954).

## II. Experimentation with Other Substances

Berger, W., u. F. J. Lang: Ein histopathologischer Beitrag zur Histaminhypothese der allergischen Reaktion. Z. Hyg. Infekt.-Kr. **113**, 206—238 (1932).

Brun, R.: Contribution à l'analyse des chromates du ciment. Helv. chim. Acta **46**, 2933—2939 (1963).

— Contribution à l'étude des chromates du ciment. Nouvelle technique pour le test épicutané au ciment. Dermatologica (Basel) **129**, 79—88 (1964).

Burckhardt, W.: Die Rolle der Alkalischädigung der Haut bei experimenteller Sensibilisierung gegen Nickel. Arch. Derm. Syph. (Berl.) **173**, 262—266 (1935).

Calnan, C. D.: Nickel sensitivity in women. Int. Arch. Allergy **11**, 73—80 (1957).

Chase, M. W.: Experimental sensitization with particular reference to picryl chloride. Int. Arch. Allergy **5**, 163—191 (1954).

Coca, A., Milford: In: Stewart, S. G., et F. E. Cormia: personal communication. J. Allergy **5**, 575—582 (1934).

Cordonnier, V.: Etudes histo-chimiques de l'intolérance cutanée dans l'eczéma expérimental. Arch. belges Derm. **5**, 321 (1949).

Dobkevitch, S., and R. L. Baer: Allergic eczematous dermatitis due to dyes in nylon stockings. Cross-sensitization to paraphenylendiamine (Preliminary report). J. invest. Derm. **8**, 419—420 (1947).

— — Eczematous cross-hypersensitivity to azodyes in nylon stockings and to paraphenylendiamine. J. invest. Derm. **9**, 203—211 (1947).

Duesberg, J. P.: La dermite allergique du cobaye; son inhibition par la cortisone et le Phénergan. Arch. belges Derm. **7**, 274—283 (1951).

— Histopathologie de la dermite allergique du cobaye, p. 70—73. In: Charpy, J. (Ed.): Le mécanisme physio-pathologique de l'eczéma. Paris: Masson 1954.

Epstein, S.: Contact dermatitis due to nickel and chromate. Observations on dermal delayed (tuberculin-type) sensitivity. Arch. Derm. (Chic.) **73**, 236—255 (1956).

— Contact dermatitis from Neomycin due to dermal delayed (tuberculin-type) sensitivity. Report of 10 cases. Dermatologica (Basel) **113**, 191—201 (1956).

— Dermal contact dermatitis. Sensitivity to Rivanol and gentian violet. Dermatologica (Basel) **117**, 287—296 (1958).

— Dermal contact dermatitis from Neomycin. Ann. Allergy **16**, 268—280 (1958).

Epstein, W. L., and A. M. Kligman: The interference phenomenon in allergic contact dermatitis. J. invest. Derm. **31**, 103—108 (1958).

da Fonseca, A.: Dermatoses pelo cromio. Contribuicao para o estudo etiopatogenico das dermites de cause externa. Porto: Imprensa Portuguesa 1963.

Graul, E. H., u. K. W. Kalkoff: Experimentelle Studien über den Vorgang der epidermalen Sensibilisierung. Arch. Derm. Syph. (Berl.) **187**, 417—430 (1948/49).

Hunziker, N.: De l'eczéma expérimental. 8ème comm. A propos de l'hypersensibilité au bichromate de potassium chez le cobaye. Dermatologica (Basel) **121**, 93—100 (1960).

— De l'eczéma expérimental. 9ème comm. Quelques expériences concernant la sensibilisation du cobaye au nickel. Dermatologica (Basel) **121**, 307—312 (1960).

— Sensibilisation du cobaye à la para-nitrosodimethylaniline (P-NDMA). A propos de l'acanthose eczémateuse. Dermatologica (Basel) **125**, 117—120 (1962).

— De l'eczéma expérimental. 20e comm. Réactions du cobaye sensibilisé à l'anhydride citraconique. Dermatologica (Basel) **129**, 89—100 (1964).

HUNZIKER, N., E. BUJARD u. W. JADASSOHN: Bemerkungen zum experimentellen Kontaktekzem des Meerschweinchens. 19. Mitt. Hautarzt **15**, 104—108 (1964).

— — — Bemerkungen zum experimentellen Kontaktekzem des Meerschweinchens. 24. Mitt. Sensibilisierung des Meerschweinchens auf Propionsäureanhydrid (PA). Arch. klin. exp. Derm. **222**, 527—532 (1965).

—, et E. MUSSO: Quelques remarques sur l'eczéma professionnel au nickel. Dermatologica (Basel) **119**, 40—45 (1959).

JACOBS, J. L.: Immediate generalized skin reactions in hypersensitive guinea pigs. Proc. Soc. exp. Biol. (N.Y.) **43**, 641—643 (1940).

— T. S. GOLDEN, and J. J. KELLEY: Immediate reactions to anhydride, of wheal-and-erythema type. Proc. Soc. exp. Biol. (N.Y.) **43**, 74—77 (1940).

JADASSOHN, W.: Eosinophile und urticarielle Quaddel. Arch. Derm. Syph. (Berl.) **166**, 458—462 (1932).

—, F. SCHAAF: Über die Häufigkeit des Vorkommens von Nickelekzem. Arch. Derm. Syph. (Berl.) **157**, 572—577 (1928/29).

KERN, R. A.: Asthma and allergic rhinitis due to sensitization to phthalic anhydride. J. Allergy **10**, 164—165 (1938/39).

LANDSTEINER, K., and J. JACOBS: Studies on the sensitization of animals with simple chemical compounds. J. exp. Med. **61**, 643—656 (1935).

— A. ROSTENBERG, and M. B. SULZBERGER: Individual differences in susceptibility to eczematous sensitization with simple chemical substances. J. invest. Derm. **2**, 25—29 (1939).

MACHER, E., u. W. SENNLAUB: Über die Sensibilisierbarkeit von Meerschweinchen durch Applikation von Dinitrochlorbenzol in Lymphknoten oder Milz. Dermatologica (Basel) **126**, 207—222 (1963).

MAGGIORA, A., E. BUJARD, and W. JADASSOHN: Different behavior of the epidermis of the flank and the nipple of the guinea pig after a 1600 R dose of X-rays. Dermatologica (Basel) **130**, 306—311 (1965).

MAYER, R. L.: Die Überempfindlichkeit gegen Körper von Chinonstruktur. Arch. Derm. Syph. (Berl.) **156**, 331—354 (1928).

— Asthma und Ekzem bei den Pelzarbeiten (Ursolasthma, Ursolekzem). Arch. Derm. Syph. (Berl.) **158**, 734—758 (1929).

— Die Ursolidiosynkrasie des Meerschweinchens. Arch. Derm. Syph. (Berl.) **163**, 223—244 (1931).

—, u. DON JACONIA: Zur Frage der Chromüberempfindlichkeit. Allergie u. Asthma **4**, 275—279 (1958).

—, u. M. B. SULZBERGER: Zur Frage der jahreszeitlichen Schwankungen der Krankheiten. Der Einfluß der Kost auf experimentelle Sensibilisierungen. Arch. Derm. Syph. (Berl.) **163**, 245—262 (1931).

MORRIS, G. E.: Chrome dermatitis. A study of the chemistry of shoe leather with particular reference to basic chromic sulfate. Arch. Derm. (Chic.) **78**, 612—618 (1958).

MUSSO, E., et R. BRUN: Essai d'un traitement désensibilisant chez des cobayes sensibilisés au blanc d'oeuf. Acta allerg. (Kbh.) **17**, 105—111 (1962).

—, N. HUNZIKER et A. MAGGIORA: Quelques remarques au sujet d'une enquête sur l'eczéma au ciment en Europe. Acta allerg. (Kbh.) **27**, 293—299 (1962).

NILZEN, A., and K. WIKSTRÖM: The influence of lauryl sulfate on the sensitization of guinea-pigs to chrome and nickel. Acta derm.-venereol. (Kbh.) **35**, 292—299 (1955).

SCHWARZ-SPECK, M., u. H. KEIL: Experimentelles Chrom-III-Ekzem. Dermatologica (Basel) **130**, 373—384 (1965).

SOLTERMANN, W.: Toxic epidermal necrolysis (Lyell.). Syndrom oder besonders schwere Verlaufsform einer Allergodermie? Dermatologica (Basel) **118**, 265—278 (1959).

STEWART, S. G., and F. E. CORMIA: Experimental nickel dermatitis. J. Allergy **5**, 575–582 (1934).

VANDENBERG, J. J., and W. L. EPSTEIN: Experimental nickel contact sensitization in man. J. invest. Derm. **41**, 413—418 (1963).

VINSON, L. J., and B. R. CHOMAN: Percutaneous absorption and surface-active agents. J. Soc. Cosm. Chem. **11**, 177—137 (1960).

WALTHARD, B.: Die Erzeugung experimenteller Nickelidiosynkrasie bei Laboratoriumstieren. Schweiz. med. Wschr. **56**, 603—604 (1926).

WIKSTRÖM, K.: Epidermal treatment of guinea pigs with potassium bichromate. Acta derm. venereol. (Kbh.) **42**, Suppl. 49, 1—59 (1962).

ZAK, F. G., M. J. FELLNER, and A. J. GELLER: Toxic epidermal necrolysis (Lyell). The scalded skin syndrome. Amer. J. Med. **37**, 140—146 (1964).

ZELIGMAN, I.: Experimental contact dermatitis. II. Contact dermatitis in guinea-pigs induced by paraphenylenediamine and related compounds. J. invest. Derm. **28**, 121—135 (1957).

## III. Hematological Observations
### (Dinitrochlorobenzene, Citraconic Anhydride and Propionic Anhydride)

HUNZIKER, N.: Experimental eczema. 23rd comm. Eosinophilic leucocytes in the guinea pig sensitized to citraconic anhydride. Acta allerg. (Kbh.) **20**, 164—170 (1965).

— R. BRUN, and W. JADASSOHN: Basophilia in sensitized guinea-pigs. Experientia (Basel) **21**, 590 (1965).

MICHELS, N. A.: The mast cells. In: Downey's Handbook of Hematology, Vol. 1, p. 232—373, 1938. Ann. N.Y. Acad. Sci. **103**, 232—373 (1963).

SEEBERG, G.: Studies of the peripheral white blood count in guinea pigs sensitized to 2:4 dinitrochlorobenzene. Acta derm. venereol. (Kbh.) **33**, 359—371 (1953).

WINQVIST, G.: Experimental production of basophil granulocytes in the guinea pig. Exp. Cell Res. **19**, 7—12 (1960).

## IV. Comparison and Discussion of the Results of the Various Sensitizations

LEVADITI, C.: Contribution à l'étude des Mastzellens et de la Mastzellen leucocytose. Thèse Paris No. 183, 1902.

MACHER, E., u. W. SENNLAUB: Über die Sensibilisierbarkeit von Meerschweinchen durch Applikation von Dinitrochlorbenzol in Lymphknoten oder Milz. Dermatologica (Basel) **126**, 207—222 (1963).

SCHLECHT, H.: Über die Einwirkung von Seruminjektionen auf die Eosinophilen und Mastzellen des menschlichen und tierischen Blutes. Dtsch. Arch. klin. Med. **98**, 308—329 (1909—1910).

SHELLEY, W. B.: The circulating basophil as an indicator of hypersensitivity in man. Arch. Derm. (Chic.) **88**, 759—767 (1963).

WINQVIST, G.: Experimental production of basophil granulocytes in the guinea pig. Exp. Cell Res. **19**, 7—12 (1960).

## V. Remarks on Passive Transfer

BAER, R. L., F. SERRI, and D. KIRMAN: Attempts at passive transfer of allergic eczematous sensitivity in man by means of white cell suspension. J. invest. Derm. **19**, 217—225 (1952).

—, M. B. SULZBERGER: Attempts at passive transfer of allergic eczematous sensitivity in man. J. invest. Derm. **18**, 53—59 (1952).

BALLESTERO, L. H., and A. H. MOM: Passive transfer of experimental contact dermatitis with the Urbach-Koenigstein technique. Ann. Allergy **3**, 435—439 (1945).

BIZZOZZERO, E., A. FERRARI: Sull'idiosincrasia all'iodoformo. G.ital. Derm. Sif. **72**, 3—24 (1931).

BRANDT, R., u. J. KONRAD: Passive Übertragung einer Kanincheneiweißidiosynkrasie mittels der Blasenmethode von Königstein-Urbach. Arch. Derm. Syph. (Berl.) **161**, 485—491 (1930).

CHASE, H. W.: The cellular transfer of cutaneous hypersensitivity to tuberculin. Proc. Soc. exp. Biol. (N.Y.) **59**, 134—135 (1945).

— Studies on the sensitization of animals with simple chemical compounds. X. Antibodies inducing immediate-type skin reactions. J. exp. Med. **86**, 489—514 (1947).

— Immunological reactions mediated through cells. In: PAPPENHEIMER JR., A. M. (Ed.): The nature and significance of the antibody response, p. 156. New York: Columbia University Press 1953.

CREPEA, S. B., and R. A. COOKE: Study on the mechanism of dermatitis venenata in the guinea pig with a demonstration of skin-sensitizing antibody by passive transfer. J. Allergy 19, 353—357 (1948).

EPSTEIN, W. L., and A. M. KLIGMAN: Transfer of allergic contact-type delayed sensitivity in man. J. invest. Derm. 28, 291—304 (1957).

FELLNER, B.: Überimpfungsversuche mit Pirketschen Papelsubstanzen am Menschen. Wien. klin. Wschr. 32, 936—941 (1919).

— Experimentelle Beiträge zum Nachweis der cellulären Sitzes der Idiosynkrasie (Veramon, Cofferylin). Klin. Wschr. 12, 540—542 (1933).

FUHS, H., u. G. RIEHL JR.: Über familiäre Salvarsanidiosynkrasie und ihre gelungene passive Übertragung. Arch. Derm. Syph. (Berl.) 154, 88—95 (1928).

GRIMMER, H., u. H. W. SPIER: Histologische Verifizierung der passiven Übertragbarkeit des tierexperimentellen Kontaktekzems. Zur Interferenz toxischer und allergischer Epicutanreaktionen. Hautarzt 12, 300—306 (1961).

GROTH, O.: Studies on contact skin reactions and normal skin of passively sensitized guinea-pigs.1. A histological study of test lesions in recipients after passive transfer of dinitrochlorobenzene (DNCB) hypersensitivity with lymphoid cells. Acta Soc. Med. upsalien. 68, 193—214 (1963).

GUTHRIES, R. K., G. E. LOWKE, J. K. FERGUSON, and W. L. ELLIS: Contact hypersensitivity to simple chemicals. Time after donor sensitization as a factor in passive transfer. J. invest. Derm. 46, 224—229 (1966).

HAGERMAN, G.: How is epidermal hypersensitivity transmitted through lymphocytes. Acta derm.-venereol. (Kbh.) 34, 51—56 (1954).

HAMILTON, L. D., and M. W. CHASE: Labelled cells in the cellular transfer of delayed hypersensitivity. Fed. Proc. 21, 40 (1962).

HARBER, L. C., and R. L. BAER: Attempts to transfer eczematous contact-type allergy with whole blood transfusions. J. invest. Derm. 36, 55—60 (1961).

HAXTHAUSEN, H.: Studies on the role of the lymphocytes as transmitter of the hypersensitiveness in allergic eczema. Acta derm.-venereol. (Kbh.) 27, 275—286 (1947).

— Passive transmission of dinitrochlorbenzene allergy with white blood cells from sensitized guinea pigs. Acta derm.-venereol. (Kbh.) 31, 659—665 (1951).

— Experiments on passive transfer of eczematous allergy. J. invest. Derm. 19, 293—296 (1952).

HUNZIKER, N.: Experimental eczema. 22nd comm. "Passive transfer" of guinea pig's hypersensitivity to citraconic anhydride. Dermatologica (Basel) 129, 475—478 (1964).

JADASSOHN, W.: Die Immunbiologie der Haut. Handb. Haut- u. Geschlechtskr., Bd. 2, S. 353—478. Berlin: Springer 1932.

JETER, W. S., M. M. TREMAINE, and P. M. SEDDOHM: Passive transfer of delayed hypersensitivity to 2,4 Dinitrochlorobenzene in guinea-pigs with leucocytes extracts. Proc. Soc. exp. Biol. (N.Y.) 86, 251—253 (1954).

KARUSH, F., and H. N. EISEN: A theory of delayed hypersensitivity. The man features of this phenomenon are explicable in terms of high-affinity humoral antibody. Science 136, 1032—1039 (1962).

KIND, P. D., F. C. BOCOBO, A. C. CURTIS, and P. BULALA: Cellular passive transfer of contact hypersensitivity to paraphenylenediamine and to 2,4-dinitrochlorobenzene in guinea pigs. J. invest. Derm. 44, 7—11 (1965).

KÖNIGSTEIN: Wien. dermatol. Ges. 17. XI. 1921.

LANDSTEINER, K., and M. CHASE: Experiments on transfer of cutaneous sensitivity to simple compounds. Proc. Soc. exp. Biol. (N.Y.) 49, 688—690 (1942).

LAWRENCE, H. S.: Delayed hypersensitivity and the behavior of the cellular transfer system in animal and man, p. 453—461. In: J. H. SHAFFER, G. A. LOGRIPPO, and M. W. CHASE (Ed.): Mechanisms of hypersensitivity. Boston-Toronto: Little Brown Co. 1959.

LEIDER, M., and R. L. BAER: The present status of passive transfer antibodies in allergic eczematous contact-type dermatitis. Failure to demonstrate passive transfer antibodies. J. invest. Derm. 10, 425—433 (1948).

McCluskey, R. T., R. Benacerraf, and J. W. McCluskey: Studies on the specificity of the cellular infiltrate in delayed hypersensitivity reactions. J. Immunol. 90, 466—477 (1963).

Meneghini, C. L., and G. Cozza: The passive transfer of the allergic contact-type sensitivity, with particular regard to the passive transfer by means of human lymphocytes. Read at VIth international Congress of Allergology Stockholm, sept. 5—9, 1965.

—, et L. Levi: Allergie eczémateuse: essais de transport passif. Acta allerg. (Kbh.) 13, 432—441 (1959).

— — et G. Cozza: Observations ultérieures sur la tentative de transport passif de la sensibilisation du type eczémateux par contact moyennant transfusions de sang, homogénats et greffes cutanées. p. 400—403. C.R. Vth European Congress of Allergology, ed. R. Schuppli. Basel: Schwabe 1963.

Metaxas, M. N., and M. Metaxas-Bühler: Passive transfer of local cutaneous hypersensitivity to tuberculin. Proc. Soc. exp. Biol. (N.Y.) 69, 163—165 (1948).

— — Über passive lokale Tuberkulinallergie. Rev. suisse path. bact. 11, 414—422 (1948).

Najarian, J. S., and J. D. Feldman: Passive transfer of tuberculin sensitivity by tritiated thymidine-labeled lymphoid cells. J. exp. Med. 114, 779—790 (1961).

— — Passive transfer of transplantation immunity. 1. Tritiated lymphoid cells. II. Lymphoid cells in millipore chambers. J. exp. Med. 115, 1083—1093 (1962).

— — Passive transfer of contact sensitivity by tritiated thymidine-labeled lymphoid cells. J. exp. Med. 117, 775—780 (1963).

— — Specificity of passively transferred delayed hypersensitivity. J. exp. Med. 118, 341—352 (1963).

Nilzen, A.: Some endocrine aspects of skin sensitization and primary irritation. 1. Observations on the influence of cortisone on cutaneous irritation and sensitization induced by various chemical compounds. J. invest. Derm. 18, 7—35 (1952).

Perutz, A.: Untersuchungen über die Jodoformdermatitis. Arch. Derm. Syph. (Berl.) 154, 206—216 (1928).

Schröpl, F., H. Röckl u. U. Grünberg: Zur Problematik der passiven Übertragung der allergischen Spätreaktion vom Ekzemtyp. Arch. klin. exp. Derm. 225, 335—352 (1966).

Serri, F.: Tentativi di trasporto passivo della sensibilita al 2:4 dinitrochlorobenzene nelle cavia a mezzo di poltiglie di fegato, polmoni, midollo, surreni e milza. G. ital. Derm. Sif. 95, 541—550 (1954).

— Sulla possibilità di trasporto passivo della sensibilità eczematosa allergica del tipo da contatto nell'uomo a mezzo di trasfusioni di sangue totale. Minerva derm. 34, 542 (1959).

Skog, E.: Experimental studies on hypersensitivity to 2,4 dinitrochlorbenzene and tuberculin in animals. Acta derm. venereol. (Kbh.) 35, 93—106 (1955).

Spier, H. W.: Zur Pathogenese des Ekzems. Fortschr. prakt. Derm. u. Ven., Bd. 5, S. 150—165. Berlin-Heidelberg-New York: Springer 1965.

Sulzberger, M. B.: Dermatologic allergy. Springfield: Thomas 1940.

Turk, J. L.: The passive transfer of delayed hypersensitivity in guinea pigs by the transfusion of isotopically labeled lymphoid cells. Immunology 5, 478—488 (1962).

—, and G. L. Asherson: Attempts to transfer contact sensitivity passively with subcellular fractions in the guinea pig. A study of the specificity of such reactions. Int. Arch. Allergy 21, 321—325 (1962).

Urbach, F.: Zur Pathogenese der cutanen Idiosynkrasien. Arch. Derm. Syph. (Berl.) 148, 146—157 (1925).

De Weck, A., et R. Brun: De l'eczéma expérimental. 4ème comm. A propos de l'histologie des réactions obtenues par transfert de la sensibilité. Acta derm.-venereol. (Stockh.) 36, 360—368 (1956).

## VI. Influence of Various Factors on Sensitization

Adatto, R.: Eczéma et rayons X. Quelques expériences chez le cobaye. Thèse Genève, No. 2838, 1962.

Baldridge, G. D., and A. M. Kligman: Preliminary and short reports. The effect of cortisone on experimentally induced contact dermatitis. J. invest. Derm. 17, 257—259 (1951).

6*

BUJARD, E., W. JADASSOHN et E. MUSSO: Cortisone et mitoses épidermiques. Schweiz. med. Wschr. **84**, 484 (1954).

FRANCESCHETTI, A., W. JADASSOHN, R. S. MACH, A. E. MASTRANGELO, J. B. BOURQUIN et J. NARDIN: Sur quelques résultats négatifs observés au cours d'expériences avec la cortisone. Infections herpétiques et vaccinales de la cornée du lapin, ophtalmoréaction tuberculinique des bovidés, réaction urticarienne dans le test de Prausnitz-Küstner, érythème cutané par rayons ultra-violets. Schweiz. med. Wschr. **81**, 924—926 (1951).

FREY, J. R., u. A. STUDER: Cortison und experimentelles Kontaktekzem mit Dinitrochlorbenzol am Meerschweinchen. Dermatologica (Basel) **103**, 65—74 (1951).

HITCH, J. M.: Action of cortisone on skin of experimental animals. Arch. Derm. (Chic.) **68**, 256—265 (1953).

HOFER, R., et M. GOLAY: Action de la cortisone sur la papule urticarienne provoquée par la morphine. Dermatologica (Basel) **116**, 197—200 (1958).

HUNZIKER, N.: De l'eczéma expérimental. 13ème comm. Prednisone, triamcinolone, G 27202 et eczéma au dinitrochlorobenzène chez le cobaye. Dermatologica (Basel) **122**, 455—459 (1961).

JADASSOHN, W., et E. BUJARD: Nouvelles expériences sur le tétine de cobaye utilisée comme organe-test. A propos de variations de la sensibilité aux rayons X. Dermatologica (Basel) **105**, 321—327 (1952).

— — R. PAILLARD, P. WENGER et P. GAUDIN: A propos de l'effet des rayons de Roentgen, des rayons $\gamma$ et des rayons $\beta$ sur la cellule épidermique. Acta radiol. (Stockh.) **34**, 469—487 (1950).

— R. S. MACH et J. NARDIN: Effet de la cortisone sur la mycose expérimentale du cobaye. Acta endocr. (Kbh.) **6**, 351—355 (1951).

KEMP, T. S., and A. M. KLIGMAN: The effect of X-rays on experimentally produced acute contact dermatitis. J. invest. Derm. **23**, 423—425 (1954).

LOZERON, H.: Diminution de la réaction eczémateuse au dinitrochlorobenzène chez les cobayes irradiés au préalable. Thèse Genève No. 2874, 1964.

MAGGIORA, A.: L'épilation par rayons X chez le cobaye. Dermatologica (Basel) **123**, 106—114 (1961).

—, et R. BRUN: Action des rayons X sur le flanc du cobaye. Atrophie — pseudoacanthose — acanthose. Dermatologica (Basel) **125**, 112—116 (1962).

— E. BUJARD, and W. JADASSOHN: Different behavior of the epidermis of the flank and the nipple of the guinea pig after a 1600 R dose of X rays. Dermatologica (Basel) **130**, 306—311 (1965).

— — — Verminderung des Meerschweinchen-Ekzems durch eine vorangegangene Röntgenbestrahlung. Derm. Wschr. **153**, 601—604 (1967).

—, et H. LOZERON: De l'eczéma expérimental. 16ème comm. Diminution de la réaction eczémateuse chez les cobayes irradiés au préalable. Dermatologica (Basel) **124**, 240—244 (1962).

MIESCHER, G., u. C. E. SONCK: Cortison und Ekzem. Bull. schweiz. Akad. med. Wiss. **8**, 188—193 (1952).

MUSSO, E.: Quelques nouveaux résultats négatifs observés au cours d'expériences avec la cortisone et l'hydrocortisone (dermite à l'huile de croton, Schick-test, dermite aux rayons ultra-violets, herpès simplex et vaccine de la cornée, poussée mitotique par unguentum cetylicum. Acta endocr. (Kbh.) **21**, 77—85 (1956).

NILZEN, A.: Some endocrine aspects of skin sensitization and primary irritation. 1. Observations on the influence of cortisone on cutaneous irritation and sensitization induced by various chemical compounds. J. invest. Derm. **18**, 7—35 (1952).

SULZBERGER, M. B., u. R. BAER: see the discussion about the publication of FREY STUDER: Cortison und experimentelles Kontaktekzem mit Dinitrochlorbenzol am Meerschweinchen. p. 382. Year Book of Dermatology and Syphilplogy 1951, Chicago: The Year Book Publ. 1952.

—, and A. ROSTENBERG: Acquired specific hypersensitivity (allergy) to simple chemicals. IV. A method of experimental sensitization; and demonstration of increased susceptibility in individuals with eczematous dermatitis of contact type. J. Immunol. **36**, 17—27 (1939).

## VII. Desensitization

BAER, R. L., S. A. ROSENTHAL, and B. HAGEL: The effect of feeding simple chemical allergens to pregnant guinea pigs upon sensibility of their offspring. J. Immunol. **80**, 429—434 (1958).

BILLINGHAM, R. E., L. BRENT, and P. B. MEDAWAR: Actively acquired tolerance of foreign cells. Nature (Lond.) **172**, 603—606 (1953).

BRENT, L.: Transplantation immunity and hypersensitivity. p. 555—568. In: J. H. SHAFFER, G. A. LOGRIPPO, and M. W. CHASE (Ed.): Boston-Toronto: Little, Brown Co. 1959.

BURCKHARDT, W.: Experimentelle Sensibilisierungen des Meerschweinchens gegen Terpentinöl (Pinen). Acta derm.-venereol. (Stockh.) **19**, 359—367 (1938).

CHASE, M. W.: Inhibition of experimental drug allergy by prior feeding of the sensitizing agent. Proc. Soc. exp. Biol. (N.Y.) **61**, 257—259 (1946).

— Studies on the mechanism of the inhibition of experimental drug allergy by prior feeding of the sensitizing agent. Bact. Proc. **1949**, 75.

—, and J. R. BATTISTO: Immunologic unresponsiveness to allergenic chemicals. In: SHAFFER, J. H., G. A. LOGRIPPO, and M. W. CHASE (Ed.): Mechanism of hypersensitivity, p. 507—517. Boston-Toronto: Little, Brown Co. 1959.

COE, J. E., and S. B. SALVIN: The specificity of allergic reactions. VI. Unresponsiveness to simple chemicals. J. exp. Med. **117**, 401—423 (1963).

FELTON, L. D.: The significance of antigen in animal tissues. J. Immunol. **61**, 107—117 (1949).

— G. KAUFFMANN, B. PRESCOTT, and B. OTTINGER: Studies on the mechanism of the immunological paralysis induced in mice by pneumococcal polysaccharides. J. Immunol. **74**, 17—26 (1955).

FREI, W.: Über willkürliche Sensibilisierung gegen chemischdefinierte Substanzen. II. Mitt. Untersuchungen mit Neosalvarsan am Tier (Salvarsanexantheme beim Tier). Klin. Wschr. **7**, 1026—1031 (1928).

FREY, J. R.: Quantitative Untersuchungen bei epidermaler Sensibilisierung von Meerschweinchen mit Dinitrochlorbenzol. Dermatologica (Basel **102**, 1—11 (1951).

—, u. H. GELEICK: Desensibilisierung durch intravenöse Injektion von Dinitrochlorbenzol beim Kontaktekzem des Meerschweinchens. Dermatologica (Basel) **125**, 132—139 (1962).

— —, and A. DE WECK: Immunological tolerance induced in animals previously sensitized to simple chemical compounds. Science **144**, 853—854 (1964).

— A. DE WECK, and H. GELEICK: Inhibition of the contact reaction to dinitrochlorobenzene by intravenous injection of dinitrobenzene sulfonate in guinea pigs sensitized to dinitrochlorobenzen. J. invest. Derm. **42**, 189—196 (1964).

FRUCHARD, J., et J. FRUCHARD: Nouvel essai de désensibilisation au bichromate de potassium dans un eczéma du ciment. Bull. Soc. franç. Derm. Syph. **64**, 776 (1967).

GINSBERG, J. E., F. T. BECKER, and S. W. BECKER: Sensitization of guinea-pigs to poison Ivy. Arch. Derm. (Chic.) **36**, 1165—1170 (1937).

GOMEZ-ORBANEJA, J., u. E. BARIENTOS: Funktionelle Nachuntersuchungen bei Ekzematikern. Schweiz. med. Wschr. **19**, 694—697 (1938).

GROLNICK, M.: Studies in contact-dermatitis. VIII. The effect of feeding of antigen on the subsequent development of skin sensitization. J. Allergy **22**, 170—174 (1951).

HARBER, L. C., S. A. ROSENTHAL, and R. L. BAER: Actively acquired tolerance to dinitrochlorobenzene. J. invest. Derm. **37**, 241—242 (1961).

HELLIER, F. F.: Modern practice in dermatology. In: G. B. MITCHELL-HEGGS, p. 142. London: Butterworth 1950.

HUNZIKER, N.: De l'eczéma expérimental. 10ème comm. Nouvelles expériences de désensibilisation dans l'eczéma expérimental. Dermatologica (Basel) **122**, 103—106 (1961).

— De l'eczéma expérimental. 11ème comm. A propos de la „désensibilisation locale" dans l'eczéma au dinitrochlorobenzène. Dermatologica (Basel) **122**, 194—199 (1961).

— Quelques remarques sur la désensibilisation dans l'eczéma expérimental du cobaye au dinitrochlorobenzène. 18ème comm. Dermatologica (Basel) **126**, 95—105 (1963).

HUNZIKER, N.: Experimental eczema. Desensitization in the experimental eczema of the guinea pig due to paranitrosodimethylaniline (PNDMA). Acta allerg. (Kbh.) 23, 83—87 (1968).

—, et E. MUSSO: A propos de l'eczéma au ciment. Dermatologica (Basel) 121, 204—212 (1960).

JADASSOHN, J.: Bemerkungen zur Sensibilisierung und Desensibilisierung bei den Ekzemen im Anschluß an einen Fall von Odolekzem. Klin. Wschr. 2, 1680—1684 (1923).

JADASSOHN, W., R. BRUN et E. BUJARD: Désensibilisation dans l'eczéma expérimental. Dermatologica (Basel) 119, 186—195 (1959).

KLIGMAN, A. M.: Poison Ivy (Rhus) dermatitis. Arch. Derm. (Chic.) 77, 148—180 (1958).

— Cashew nut shell oil for hyposensitization against Rhus dermatitis. Arch. Derm. (Chic.) 78, 359—363 (1958).

— Fundamental of modern allergy. S. J. PRIGAL, ed., p. 322—357. New York: McGraw Hill Book Co. Inc. 1960.

LOWNEY, E. D.: Topical hyposensitization of allergic contact sensitivity in the guinea pig. J. invest. Derm. 43, 487—490 (1964).

— Immunologic unresponsiveness appearing after topical application of contact sensitizers to the guinea pig. J. Immunol. 95, 397—403 (1965).

MAGUIRE, H. C., and H. I. MAIBACH: Effect of cyclophosphamide, 6-mercaptopurine, actinomycin D and vincaleukoblastine on the acquisition of delayed hypersensitivity (DNCB contact dermatitis) in the guinea pig. J. invest. Derm. 37, 427—431 (1961).

— — Effect of X-ray lymphopenia on contact dermatitis. Arch. Derm. (hic.) 88, 768—770 (1963).

MAIBACH, H. I., and H. C. MAGUIRE: Elicitation of delayed hypersensitivity (DNCB contact dermatitis) in markedly panleukopenic guinea pigs. J. invest. Derm. 41, 123—127 (1963).

MIESCHER, G.: Die Bedeutung der Testproben für die Haut. Arch. Derm. Syph. (Berl.) 177, 8—35 (1938).

MORGAN, J. K.: Observations on the persistence of skin sensitivity with reference to nickel eczema. Brit. J. Derm. 65, 84—94 (1953).

RAJKA, G., and S. HARD: Changes in guinea-pig skin resulting from repeated application of dinitrochlorbenzene at the same site. Acta derm. venereol. (Kbh.) 40, 64—73, (1960).

ROSENTHAL, S. A., and R. L. BAER: Actively acquired tolerance to dinitrochlorobenzene tests in newborn guinea pigs. J. invest. Derm. 41, 351—355 (1963).

SCHIMPF, A., u. G. FILIPP: Untersuchungen zur Frage der Immunotoleranz gegenüber Dinitrochlorbenzol. Acta allerg. (Kbh.) 20, 187—196 (1965).

SCHULZ, K. H.: Experimentelle Untersuchungen zur Desensibilisierung von allergischen Kontaktekzemen. In: LETTERER, E., u. W. GRONEMEYER (Ed.): Allergie und Immunitätsforschung, S. 143—153. Stuttgart: Schattauer 1965.

SEQUEIRA, J. H., J. T. INGRAM, and R. T. BRAIN: Diseases of the skin. 5th ed., quoted by MORGAN, ed. London: J. & A. Churchill 1947.

SOURREIL, P., J. FRUCHARD et J. FRUCHARD: Désensibilisation dans un eczéma du ciment (2ème présentation). Bull. Soc. franc. Derm. Syph. 71, 751 (1964).

— — — Nouvelle désensibilisation dans l'eczéma du ciment. Bull. Soc. franc. Derm. Syph. 71, 752 (1964).

SULZBERGER, M. B.: Hypersensitiveness to neoarsphenamine in guinea pigs. 1. Experiments in prevention and in desensitization. Arch. Derm. (Chic.) 20, 669—697 (1929).

— Zur Frage der experimentellen Salvarsan-Überempfindlichkeit. Klin. Wschr. 9, 253—254 (1929).

TEES, E. C., and F. H. MILNER: Hyposensitization in nickel dermatitis. Acta allerg. (Kbh.) 15, suppl. VII, 413—418 (1960).

DE WECK, A. L., and J. R. FREY: Monographs in allergy. Immunotolerance to simple chemicals. Basel-New York: S. Karger 1966.

WEDROFF, N. S., u. A. P. DOLGOFF: Über die spezifische Sensibilität der Haut einfachen chemischen Stoffen gegenüber. Arch. Derm. Syph. (Berl.) 171, 647—664 (1935).

Weigle, W. D., and F. J. Dixon: Immunologic unresponsiveness to protein antigen. In: Shaffer, J. H., G. A. LoGrippo, and M. W. Chase (Ed.): Mechanisms of hypersensitivity. 529—536. Boston-Toronto: Little, Brown Co. 1959.

White, W. A., and R. L. Baer: Failure to prevent experimental eczematous sensitization; observations on the "spontaneous" flare-up phenomenon. J. Allergy **21**, 344—348 (1950).

Zagula, Z. W. J., H. C. Maguire, and H. I. Maibach: Dermatitis in leukopenic guinea pigs. J. invest. Derm. **41**, 405—411 (1963).

## VIII. Discussion of Some Questions. IX. Conclusion

Baer, R. L., S. A. Rosenthal, and Ch. Sims: Contact dermatitis with spongiosis and intraepidermal vesiculation in the acanthotic skin of guinea pigs. J. invest. Derm. **27**, 249—258 (1956).

Bandmann, H. J.: Beitrag zur Histopathologie allergischer epicutaner Testreaktionen. Besprechung der eigenen Untersuchungsergebnisse. Hautarzt **11**, 393—400 (1960).

Bloch, Br.: Ekzem und Überempfindlichkeit. Schweiz. med. Wschr. **53**, 629—630 (1923).

—, u. A. Steiner-Wourlisch: Die Sensibilisierung des Meerschweinchens gegen Primeln. Arch. Derm. Syph. (Berl.) **162**, 349—378 (1930).

Brun, R., W. Jadassohn et R. Paillard: Test épicutané avec réaction immédiate (persulfate d'ammonium). Dermatologica (Basel) **133**, 89—90 (1966).

Brunsting, L., and R. L. Bailey: Ragweed (contact) dermatitis produced experimentally in the guinea pigs. J. Allergy **6**, 547—550 (1935).

Calnan, C. D., and S. Shuster: Reactions to ammonium persulfate. Arch. Derm. Syph. (Chic.) **88**, 812—815 (1963).

Chase, M. W.: Inheritance in guinea pigs in the susceptibility to skin sensitization with simple chemical compounds. J. exp. Med. **73**, 711—726 (1941).

Civatte, A.: Eczéma et eczématides. Bull. Soc. franç. Derm. Syph. **32**, 134—147 (1925).

— Anatomie pathologique de l'eczéma. Bull. Soc. franç. Derm. Syph. **57**, 13—32 (1950).

— La formule histologique de l'eczéma. In: Charpy, J. (Ed.): Le mécanisme physiopathologique de l'eczéma, p. 55—60. Paris: Masson 1954.

Epstein, S.: The antigen-antibody reactions in contact dermatitis. Ann. Allergy **10**, 633—658 (1952).

— Contact dermatitis from Neomycin due to dermal delayed (tuberculin-type) sensitivity. Report of 10 cases. Dermatologica (Basel) **113**, 191—201 (1956).

— Contact dermatitis due to nickel and chromate. Observations on dermal delayed (tuberculin-type) sensitivity. Arch. Derm. (Chic.) **73**, 236—280 (1958).

— Dermal contact dermatitis from Neomycin. Ann. Allergy **16**, 268—280 (1958).

— Dermal contact dermatitis. Sensitivity to Rivanol and gentian violet. Dermatologica (Basel) **117**, 287—296 (1958).

Epstein, W. L., and A. M. Kligman: Some factors affecting the reaction of allergic contact dermatitis. J. invest. Derm. **33**, 231—243 (1959).

Ginsberg, J., F. Becker, and W. Becker: Sensitization of guinea-pigs to poison. Ivy. Arch. Derm. (Chic.) **36**, 1165—1170 (1937).

Ginsberg, J. E., C. D. Stewart, and S. W. Becker: Cutaneous sensitization studies. II. Cross and microscopic changes in ragweed and 2-4 dinitrochlorobenzene sensitization of guinea pigs, and in poison Ivy sensitization of human beings. J. Invest. Derm. **2**, 81—92 (1939).

de Graciansky, P., L. Israel et J. Cohen-Solal: L'eczéma du cobaye au dinitrochlorobenzène. Etude de l'influence de certains médicaments neurotropes. Sem. Hôp. Paris **36**, 1451/S P-303—1457/S P-309 (1960).

— — — L'eczéma du cobaye au dinitrochlorobenzène. II. Détermination quantitative de la sensibilité. Notion de seuil. Individualité des cobayes. Sem. Hôp. Paris **36**, 1458/SP-310—1460/SP-312 (1960).

— — — L'eczéma du cobaye au dinitrochlorobenzène. III. L'influence de certains médicaments neurotropes sur l'eczéma expérimental du cobaye. Sem. Hôp. Paris **36**, 1461/SP-313—1464/SP-316 (1960).

GRIMMER, H.: Die Histologie der epicutanen Testreaktion bei nicht epicutaner („extra-cutaner") Sensibilisierung mit 2,4-Dinitrochlorbenzol. Arch. klin. exp. Derm. **214**, 105—113 (1961).

—, u. H. W. SPIER: Histologische Verifizierung der passiven Übertragbarkeit des tier-experimentellen Kontaktekzems. Zur Interferenz toxischer und allergischer Epicutan-reaktionen. Hautarzt **12**, 300—306 (1961).

HAXTHAUSEN, H.: Allergy in diseases of the skin (p. 183). In: KALLOS, P.: Progress in allergy, Vol. 2, p. 167—235. Basel-New York: S. Karger 1949.

HUNZIKER, N., E. BUJARD u. W. JADASSOHN: Bemerkungen zum experimentellen Kontakt-ekzem des Meerschweinchens. 19. Mitt. Hautarzt **15**, 104—108 (1964).

JADASSOHN, W.: Sensibilisierung der Haut des Meerschweinchens auf Phenylhydrazin. Klin. Wschr. **9**, 551—552 (1930).

— Die Immunbiologie der Haut. Handb. Haut- u. Geschlechtskr. Bd. 2, S. 353—478. Berlin: Springer 1932.

— Aphoristische Bemerkungen zum Berufsekzem. Proc. Intern. congress on occupational Health, Wien 19—24. 9. 1966, 187—191 (S. 187).

— E. BUJARD, and R. BRUN: The experimental eczema of the guinea-pig nipple. J. invest. Derm. **24**, 247—253 (1955).

KLASCHKA, F.: Tierexperimentelle Studien zum Kontaktekzem. II. Histologie der epi-cutanen 2,4-Dinitrochlorbenzolhautreaktion bei „cutan" und „extracutan" sensibili-sierten Meerschweinchen. Arch. klin. exp. Derm. **224**, 216—234 (1966).

MIESCHER, G.: Zur Histologie der ekzematösen Kontaktreaktion. Dermatologica (Basel) **104**, 215—220 (1952).

— Ekzem, Histopathologie, Morphologie, Nosologie. Handb. Haut- u. Geschlechtskr. Erg. II/1, S. 1—111 (S. 40). Berlin-Göttingen-Heidelberg: Springer 1962.

PILLSBURY, D. M.: see the discussion in the publication of R. L. BAER, S. A. ROSENTHAL and CH. F. SIMS: The allergic eczema-like reaction and the primary irritant reaction. Arch. Derm. (Chic.) **76**, 549—560 (1957).

POLAK, M., and A. M. MOM: Histopathology of experimental eczema (allergic contact-type eczematous dermatitis) in man. A study by the technics of silver impregnations of Rio Hortega with special reference to the early microscopic lesions. J. invest. Derm. **13**, 125—134 (1949).

SATO, W.: Studies on epidermal sensitization. Report I. Problems of skin penetration. Nihon Univ. J. Med. **3**, 63—69 (1961).

— Studies on epidermal sensitization. Report II. Experimental studies in guinea-pigs on allergic epidermitis. Nihon Univ. J. Med. **3**, 71—78 (1961).

DE WECK, A., et R. BRUN: De l'eczéma expérimental. 4ème comm. A propos de l'histologie des réactions obtenues par transfert de la sensibilité. Acta derm. venereol. (Stockh.) **36**, 360—368 (1956).

WEDROFF, N. S., u. A. P. DOLGOFF: Ein Beitrag zur Frage der Reaktionsfähigkeit der Haut bei „chemischen" Ekzemen, auf Grund von Untersuchungen bei Ekzemkranken und gesun-den Personen mittels der obligaten chemischen Reizmittel. Arch. Derm. Syph. (Berl.) **171**, 641—646 (1935).

# Author Index

Page numbers in *italics* refer to the literature

# Subject Index